THE PERFECT FENCE

Connecting the Greater West

THE PERFECT FENCE

Untangling the Meanings of Barbed Wire

Lyn Ellen Bennett AND **Scott Abbott**

FOREWORD BY Sterling Evans

Texas A&M University Press

College Station

First edition

This paper meets the requirements of ANSI/NISO Z39.48–1992 (Permanence of Paper).
Binding materials have been chosen for durability.
Manufactured in the United States of America

Library of Congress Cataloging-in-Publication Data
Names: Bennett, Lyn Ellen, author. | Abbott, Scott H., author.
Title: The perfect fence: untangling the meanings of barbed wire / Lyn Ellen Bennett and Scott Abbott; foreword by Sterling Evans.
Description: First edition. | College Station: Texas A&M University Press, [2017] | Series: Connecting the greater west series | Includes bibliographical references and index.
Identifiers: LCCN 2017013967 (print) | LCCN 2017017043 (ebook) | ISBN 9781623495831 (ebook) | ISBN 9781623495824 | ISBN 9781623495824 (cloth : alk. paper)
Subjects: LCSH: West (U.S.)—Social life and customs—19th century. | West (U.S.)—History—1860-1890. | Wire fencing—United States—History—19th century. | Barbed wire—History—19th century. | Metaphor in literature. | American fiction—19th century—Themes, motives.
Classification: LCC F594 (ebook) | LCC F594 .B464 2017 (print) | DDC 978/.02—dc23
LC record available at https://lccn.loc.gov/2017013967

Once in a Blue Moon
Rare, Loved, Missed

CONTENTS

ILLUSTRATIONS

FOREWORD

The past couple of years (2016–2017) have been characterized by a great deal of discussion about fences and walls in US political culture. Although the Berlin Wall came crumbling down after the fall of the Soviet Union in 1991, other walls and fences separating peoples and places have been fortified or constructed, most notably along the borders of North and South Korea, Israel and Palestine, and the United States and Mexico. In fact, the Secure Fence Act of 2006, passed during the George W. Bush administration (and supported by Republicans and Democrats alike), was the legislative impetus for the construction of a fortified wall. More than a fence, 700 miles are constructed already along various segments of the international border in Texas and other parts of the Southwest. The campaign and election of President Donald Trump in 2016 accelerated that process, with talk of a much bigger and more expensive wall, with more vocalized support for and opposition against it, and with curiously far more media attention and political activism than Bush's current border "fence" ever received.

The issues regarding the development of walls and fences along international borders share common trajectories and histories, including issues such as immigration control, commercial trade, contraband and smuggling, xenophobia, and national security (especially efforts to thwart terrorist activities). All of these issues revolve around more secure borders and restricting or controlling the movement of people and goods. For the US-Mexico border wall, there is a uniquely American West element in its development, both in the location of its construction and in the inherent questions of immigration, labor, and social issues that are extremely pertinent to western economies and politics.

But the West has a long history with fences. Lyn Ellen Bennett and Scott Abbott explore this history and culture in the region deeply in *The Perfect Fence: Untangling the Meanings of Barbed Wire*, which traces the invention of barbed wire and the history of fencing the West. Although a few other books have examined this story, Bennett and Abbott add needed analysis on the material and popular culture of barbed wire using an impressive array of illustrations that exemplify this cultural history. Their work pushes further than any other study into the legislative history of barbed wire with in-depth coverage of inventions, patents, and congressional debates over use and control of the wire used across the nation. Included is a sorely needed discussion of the effects of barbed wire on Native Americans in the West. Critically, they draw barbed wire connections from the West to other regions of the United States and Europe and its wartime uses (trench warfare and Japanese internment, among other things) by analyzing the commodity through literary and cultural lenses.

The Perfect Fence is the eighth and most recent addition to the Connecting the Greater West Series, a collection of books dedicated to exploring the changing and growing ways that historians and other scholars are coming to view "the West"—a region including the American West, western Canada, northern Mexico, parts of the Pacific Rim, and the borderlands between these regions. As such, the series includes books on transnational history, borders and borderlands, immigration, agriculture, Indigenous negotiations of bordered regions in the West, environmental change, and material and popular culture of the greater region—the particular theme of *The Perfect Fence*. As you read here, you will get a sense of business history and material culture, but you will also discover important aspects of American Indian, agricultural, and transnational history. You will start seeing all of these connections and how they are tied together, and you will clearly discern the expanded notion of how we must really view the Greater West.

STERLING EVANS
Series Editor

PREFACE

We live in a small town tucked into the foothills of Utah's Wasatch Mountains. The town borders on a working ranch with alfalfa fields, grazing land, and cattle. The ranch is enclosed by a barbed wire fence: four strands of wire attached to steel posts for the most part, with cedar posts where the fence needs extra support. Sturdy cedar corner posts anchor the fence where the property line makes right-angled turns. The fence keeps the cattle in; mule deer and elk leap over and duck through it with ease. A five-wire fence around neighboring property is not as easily traversed, and several years ago a doe was caught in the fence and set free only by cutting the wire. A steep hillside bordered by the ranch's fence was the location of daily walks with our dog Blue almost every day for his twelve years. Sagebrush and scrub oak are abundant there, and in the spring the hill is alive with pink spring beauties, yellow glacier lilies (Blue preferred the name "dogtooth violets"), Wasatch bluebells, and a steady succession of wildflowers as spring gave way to summer. Blue sampled scents as we paid attention to colors, to the birds, and to the valley below us. At the fence we turned along the hill and followed the bristling wires along a deer path. Blue sniffed matted grass where deer had bedded down for the night and nosed scat of uncertain origin while we admired the morning or evening light on the twisted strands of barbed wire.

Barbed wire fences are ubiquitous in the American West, so commonplace that we had always accepted them as matters of fact rather than as matters worthy of study. What can be said, after all, about the countless miles of wires twisted around barbs and stretched between posts to control the movement of cattle and horses? It was the 2009 call for papers from the Western History Association on the theme "Wired West" that made us wonder if there might not be something about barbed wire that a historian and a literary critic might find interesting. The idea of the wired West brought to mind our friend Brian Evenson's "Contagion," a story about men hired to ride and report on a barbed wire fence who find an inexplicable contagion and a sect of people for whom the barbed fence has become an object of worship. The story slips back and forth between actual barbed wire and barbed wire as metaphor, a fascinating construction of meaning we discussed at length. Perhaps this mutability could offer new perspectives on an old technology.

When barbed wire was first conceived, we realized, it had no meaning beyond the quest to make something that would control livestock. To sell the new fence, meanings had to be constructed, and early advertisements did exactly that. Barbed wire was more than a vicious fence, we discovered; it was a meaningful thing. We decided to pursue those meanings both historically and literarily. The rest is (our) history.

ACKNOWLEDGMENTS

Many archivists, curators, conservators, and librarians have made our research possible. We thank Ginny Kilander and the staff at the American Heritage Center of the University of Wyoming for their assistance during multiple visits. It was our work there with the Frances and Henry McCallum papers, the Henry Putnam papers, the Earl Hayter Collection, and the Ellwood Family Collection that started us on this journey. Thanks to Russ Taylor and Cindy Brightenburg at Brigham Young University's Harold B. Lee Library (which houses a copy of the first barbed wire advertisement) and to Scott Reilly at the Vermont State Archives and Records Administration. Thanks to Brad Penka at the Kansas Barbed Wire Museum for help with their collection of barbed wire patent medicine bottles and to staff of the Devil's Rope Museum for answering our queries. Thanks also to the staff at the Western History/Genealogy Department of the Denver Public Library, of the Regional History Center/University Archives of the Northern Illinois University Libraries, and of the California History Section of the California State Library for their professional support as we gathered the documents discussed here. The time we spent at Harvard University's Baker Library was especially fruitful as we worked through the extensive American Steel & Wire Company collection.

A special thanks to Brian Reis, executive director of the Ellwood House Museum in DeKalb, Illinois, for his time and expertise. Stephen Bigolin was a knowledgeable guide to the sites of the barbed wire industry in DeKalb. Anonymous readers/reviewers for *Agricultural History*, and the *Steinbeck Review* (where versions of chapters 3 and 4 were originally published), and Texas A&M University Press asked well-founded questions and from the perspectives of their disciplines made invaluable suggestions.

Many thanks to Karen Sturtevant, Jacques Demal, and Catherine McIntyre at Utah Valley University Library for their sustained assistance. Thanks to our colleagues Jan Wellington for her assistance as we navigated Native American literature and followed Oscar Wilde's visit to the United States; Laura Hamblin for bringing our attention to Amy Clampitt's poem "Woodlot"; and Cheryl Hanewicz, who continues graciously to set aside some of her own research agenda to take up multiple barbed wire projects in support of Lyn's conference panels. Utah Valley University; the UVU College of Humanities, Arts, and Social Sciences; and the Departments of History and Integrated Studies have generously supported our research and travel over these seven years.

We are especially grateful to Sterling Evans, who has followed, encouraged, and contributed to this project from the original conference paper to its inclusion in his series Connecting the Greater West. It has been a pleasure working with the professional and helpful staff of Texas A&M University Press. Editor in chief Jay Dew, managing editor Katie Duelm, and copyeditor Cynthia Lindlof were exacting and flexible—qualities for which we are grateful.

Special thanks to our families, who have patiently taken this ride with us; to our friends and to our Woodland Hills neighbors for tolerating years of barbed wire conversation; and to Michael Bennett and Marina Nelson for their care of our four-legged dependents when we traveled and for the use of their cactus garden during final editing. Blue, Calico, and Bella have kept us sane and reminded us daily of how wonderful it is to live in the West.

Introduction

Like the rest of the things people make, walls both reflect and create values.
—Thomas Oles, *Walls*

In 1894, Jack London traveled with General Charles Kelly's hobo army by train and boat as it made its way across the continent toward Washington, DC. Floating down the Des Moines River, the eighteen-year-old London was part of an advance team scavenging for food. Word of their raiding preceded them, and farmers who had earlier supported them, London wrote,

> summoned the constables when we tied up to the bank, and loosed the dogs. I know. Two of the latter caught me with a barbed-wire fence between me and the river. I was carrying two buckets of milk for the pale Vienna. I didn't damage the fence any; but we drank plebeian coffee boiled with vulgar water, and it was up to me to throw my feet for another pair of trousers. I wonder, gentle reader, if you ever essayed hastily to climb a barbed-wire fence with a bucket of milk in each hand. Ever since that day I have had a prejudice against barbed wire, and I have gathered statistics on the subject.[1]

We too have gathered statistics on the subject of barbed wire, statistics and much more.

The moment barbed wire entered language—in patents, public debates, and early advertising—it began to take on an array of meanings. The qualities of barbed wire were intensely debated in American newspapers. The economical and effective fence facilitated European American settlement of the American West and in doing so became a contested symbol of conquest. Creative advertisers sought ways to make a desperately needed but dangerous fence attractive to customers concerned about the well-being of their

Figure 1. A Google Ngram search for "barbed wire" shows a general rise in use of the term since its invention, with distinct upturns during the two world wars.

animals. And meanings proliferated when barbed wire became indispensable on African and European battlefields, in British concentration camps during the Boer War, and then in the concentration camps of the Holocaust.

A Google Ngram search reveals a steady rise in references to barbed wire in the corpus of available books (fig. 1).[2] Beginning at a zero point in 1874 (a critical year for invention and manufacture of barbed wire), references grow more and more common as the wire comes into wide use, and there are dramatic increases during the world wars. A quick Internet search for barbed wire shows that it is much more than a fence. Along with barbed wire wreaths and flatware and sculptures, one finds song lyrics and tattoos and a logo for Amnesty International. There is a naked Russian artist wrapped in barbed wire, the singer Rihanna wrapped in nothing but barbed wire, and an Israeli artist making a political statement with a barbed wire hula hoop ripping into her naked torso. Harpo Marx plays Salvador Dalí's harp strung with barbed wire, and musicians make music with a barbed wire fence between the United States and Mexico. Barbed wire accoutrements for professional wrestling stand alongside barbed wire boots and bras and hats. There are even images of horses and cattle in fields enclosed by barbed wire. The potent tropes first exploited by advertisers and evident in this quick sampling—control exerted through threatened violence being the central one—are the focus of this book, which explores the identities barbed wire imposes and invites. The fence Jack London encountered has become a cultural icon.

The propensity for the physical fence to become metaphor is nicely illustrated by a ranch sign marking the flats where Fool Creek exits the mountains of the Canyon Range near Oak City, Utah, a site initially proposed for the atomic weapons laboratory of the Manhattan Project (fig. 2). The Great Basin

Figure 2. Ranch sign near Oak City, Utah. Photograph by the authors.

is high-desert country; ranching there is a risky and difficult enterprise. Given the sparse vegetation, large tracts of land are necessary for grazing, making barbed wire fences indispensable. The sign's calf, horse, and rope-waving cowboy are cut out of steel plate, and the pipes, the FOOL CREEK FLAT sign, and the chains are steel as well. The ring surrounding the figures is made up of multiple strands of twisted steel wire—strong, stiff wire armed with sharp barbs. This rancher, the sign indicates, is tough as steel. He (or she?) is as tautly coiled and potentially dangerous as the surrounding barbs. If anyone can scratch a living out of this land, it will be this rancher with a steel-and-barbed-wire identity. One leg of the sign performs double duty: as part of the meaningful sign and as a sturdy fence post. The material reality of the fence and the metaphors it embodies are not easily separated.

The question of American identity as related to barbed wire is best answered in a larger context. Two issues are especially interesting in this regard. One involves the shift in nineteenth-century America from a pastoral society to an industrial one. The second concerns the transfer of shared land or public property to private ownership via fence building.

In *The Machine in the Garden*, Leo Marx argues that the pastoral ideal—a withdrawal from the world for a new beginning in a bucolic setting—was a component of European American identity from the beginning.[3] That ideal is increasingly interrupted, he writes, by intruding machines. The conflict is between Thomas Jefferson's agrarian ideal (expressed in his *Notes on Virginia* [1785] in which America would retain its vigor and virtue only as it remained an agriculture society) and the ideal of industrialism that came to dominate nineteenth-century America. "The official American ideology of industrialism," Marx writes, was "a loosely composed scheme of meaning and value so widely accepted that it seldom required precise formulation. In the written record it appears chiefly as rhetoric in homage to 'progress.'" The steam engine was proof of such progress, and in the 1830s the locomotive becomes "a kind of national obsession. It is the embodiment of the age, an instrument of power, speed, noise, fire, iron, smoke." Leading magazines, Marx writes, "adduce the power of machines (steam engines, factories, railroads, and, after 1844, the telegraph) as the conclusive sanction for faith in the unceasing progress of mankind." Through what Marx calls the "rhetoric of the technological sublime," the machine becomes "a transcendent symbol: a physical object invested with political and metaphysical ideality."[4]

The industrial ideal had detractors. Marx quotes lines from Ralph Waldo Emerson's 1847 "Ode, Inscribed to W. H. Channing" as criticism of the alienation and detachment occasioned by industry:

> Things are in the saddle,
> And ride mankind.
>
> There are two laws discrete
> Not reconciled,
> Law for man, and law for thing;
> The last builds town and fleet,
> But it runs wild,
> And doth the man unking.[5]

Do we use the barbed fence, or does it use us? Emerson wrestles here with the paradox of a pastoral/industrial society. Marx traces the conflicting yet intertwined ideals as they appear in texts by Henry David Thoreau, Herman Melville, Walt Whitman, Mark Twain, and others, fascinated by the metamorphosis of the thing into a thing of meaning: "the symbolic properties of the machine image," for instance, "its capacity to embrace a whole spectrum of meanings ranging from a specific class of objects at one end to an abstract metaphor of value at the other."[6] In the chapters that follow, we examine just this spectrum of meanings for barbed wire.

The second strand of history interesting for our purposes begins with the fact that, for better or worse, barbed wire fences share the characteristics of other kinds of enclosures. Rousseau's *Discourse on Inequality, No 2* (1755) situates the origin of inequality in the first enclosure:

> The first person who, having enclosed a plot of ground, thought of saying *this is mine* and found people simple enough to believe him was the true founder of civil society. What crimes, wars, murders, what miseries and horrors, would the human race have been spared by someone who, pulling up the stakes or filling in the ditch, had cried out to his fellow humans: "Beware of listening to this imposter. You are lost if you forget that the fruits are everyone's and the earth is no one's!"[7]

Rousseau's negative assessment of enclosure finds a delightful echo in H. G. Wells's satirical *The Wonderful Visit* (1895). The novel follows a beautiful young Angel surprised to find himself on earth and even more surprised when the Vicar mistakes him for a bird and shoots him. The good Vicar helps the Angel recover and introduces him to assorted humans whose behaviors related to class and property and violence bewilder him. The Angel's naïveté eventually gets him in trouble with Sir John Gotch, a wealthy, fence-building neighbor. The narrator describes a little girl gathering flowers in a bucolic scene marred by a barbed wire fence meant to protect the "dignity" of the landowner and to keep away the detested rabble. "Then suddenly a gashed hand, a bitter outcry, and the Angel sympathetic, comforting, inquisitive. . . . A furious onslaught upon the barbed wire of Sir John Gotch, barbed wire recklessly handled, slashed, bent and broken."[8]

The dignified landowner comes to protest:

> "This business," said the Vicar, "this unfortunate business of the barbed wire—is really, you know, a most unfortunate business."
>
> Sir John Gotch became decidedly more angular in his attitude. "It is," he said.
>
> "This Mr. Angel being my guest—"
>
> "No reason why he should cut my wire," said Sir John Gotch, briefly. . . .

> "Do you know," said Sir John, with scarcely a pause, "he's been going about this village preaching Socialism? . . . He has done this wire-cutting as a demonstration, I tell you, as a Socialistic demonstration." . . .
>
> "A Socialist," said the Vicar, quite put out, "I had *no* idea."[9]

Thinking about the kinds of hierarchies established by fences, walls, and borders, landscape architect Thomas Oles reasons that walls are not inherently good or bad. His analysis of Robert Frost's poem "Mending Wall" points out that although the poem ends with the famous phrase "Good fences make good neighbors," it begins with the antithetical phrase "Something there is that doesn't love a wall." Together the two lines encompass the poem's range of meaning. In a chapter titled "Constructions of Sovereignty," Oles places the fencing of the American West in the context of the English Enclosure Acts. He paraphrases John Locke on property: "Constructing a fence or wall around a plot of land did more than merely symbolize property: it *created* it." Enclosure of the English countryside, of lands that had been commons, "transferred control over land and resources to an ever-smaller group of people." In this spirit, Europeans like John Winthrop, the first governor of the Massachusetts Bay Colony, claimed that because the Indians "enclose no Land, neither have any settled habitation, nor any tame Cattle to improve the Land by, . . . we may lawfully take the rest." "The act of enclosing land," Oles points out, "provided both the *justification* and the *instrument*" of the takeover. American enclosures repeated the "brazen theft of common resources" with the theft of lands inhabited by Native Americans.[10] A barbed wire company publication proclaiming "Steel Barb Fencing" as "The World's Fence" and "The Perfect Fence" made this explicit, contrasting civilized fence builders with nomadic savages:

> Every man who builds a fence, does so, primarily, for his own greater enjoyment in his own lands, and the sense of better security in their exclusive possession. It enables him to protect his own improvements. In no part of the world, where the people have risen above the condition of the wandering savage, does the benefit of fencing fail to be understood and appreciated.[11]

We investigate connections between barbed wire fences and Native Americans at some length in chapters 3 and 7.

The 1862 Homestead Act awarded free title to settlers who inhabited and improved the land for five years. Because the stone and timber fences that might have served as a means of improvement were largely absent, barbed wire was a welcome alternative. As settlers increasingly claimed and fenced what had been free range, big ranchers countered by enclosing huge sections of the American West with barbed wire. This takeover of public lands soon came to national attention. In February 1884, the US Senate directed Interior Secretary Henry Teller to report on the "unauthorized fencing of the public lands." In the report issued a month later, the acting commissioner of the Land Office, L. Harrison, noted that investigations

> show the general fact of the existence, upon a large scale, and to an unknown extent, of unauthorized fencing of public lands. . . . The reports . . . also show that fraudulent entries of public land within the inclosures are extensively made by the procurement and in the interest of stockmen, largely for the purpose of controlling the sources of water supply.[12]

Harrison further noted that "millions of acres are thus inclosed and are now being so inclosed to the exclusion of the stock of all others than the fence owners, and to the prevention of settlements and the obstruction of public travel and intercourse."[13] Experiences of General Land Office agents were included in the report. James Tullis, writing about conditions in Montana, noted that "there is a constant and well-grounded complaint in all western country in regard to the manner in which large stock growers control the public lands for their own use and benefit, to the exclusion of their less wealthy neighbors and small stock grazers."[14] Henry De Ahna reported that his arrival in Carson City, Nevada, stopped some farmers from completely fencing in a Shoshone Indian tribe who were attempting to raise stock near Star Valley.[15] Congress finally responded to the rampant illegal fencing with a law in February 1885 that forbade unauthorized enclosure of public lands.[16] This act was reinforced in August of that year when President Grover Cleveland issued Proclamation 271, which required the immediate removal of "any and every unlawful inclosure of the public lands."[17]

In these historical contexts, our interdisciplinary study explores relationships between barbed wire and American identity. What has barbed wire come to mean? we ask. How were those meanings established? And what effect do those meanings have on us? The first half of the book explores meanings constructed for the barbed wire fence in the final quarter of the nineteenth century. The introduction to part 1 presents the three men whose inventions turned their hometown, DeKalb, Illinois, into a manufacturing hub that for several decades fed the country's and world's voracious appetite for barbed wire. Along with some thoughts on material history, the introduction also surveys the work done by historians of barbed wire. Chapter 1 focuses on the vigorous debate that unfolds in dozens of newspapers and magazines as the controversial new fence comes into widespread use. Chapter 2 examines legislative attempts to ban or regulate barbed wire fences and responses by manufacturers to defend their product and to spin how it is perceived. We also trace how the outrage over barbed wire leads to an explosion of new patents meant to ameliorate the perceived problems. Chapter 3 analyzes dozens of sometimes outrageous advertisements meant to construct a positive sense for a dangerous but highly effective fence.

The book's second half is literary-historical in nature, focused on barbed wire as a literary motif. Chapter 4 looks at barbed wire in the traditional Western, in most cases involving a contest between advocates of free range and fence-building settlers. Chapter 5 investigates literature of the New West in which the contest shifts to a battle between ranchers and environmentalists and to the simple fact of barbed wire as a symbol of Western Identity. Chapter 6 explores works of literature in which barbed wire takes on religious significance. And the novels that are the focus of chapter 7 depict barbed wire at the heart of conflicts between Native and European Americans from the nineteenth century to the present day, with Native identity as a particular concern.

THE PERFECT FENCE

PART ONE

Constructing the Meaning of Barbed Wire in Late Nineteenth-Century America

The character of a man may be known by a glance at his surroundings.
If the farmer's buildings be snug, . . . his stock sleek, well fed and kept in their
proper places by good fences, you may be morally certain he is a
prosperous, well-to-do and influential man in his community.
You may also be certain that he used the Glidden wire.
—*Glidden Barb-Fence Journal*, 1887

Objects made or modified by humans, consciously or unconsciously,
directly or indirectly, reflect the belief patterns of individuals who made,
commissioned, purchased, or used them, and, by extension,
the belief patterns of the larger society of which they are part.
—Thomas Schlereth, *Material Culture Studies in America*

As you drive west on Interstate 88 from Chicago, the tangled streets of the metropolis give way to the geometrically ordered fields that dominate this midwestern landscape. Sixty miles east of Chicago lies DeKalb, Illinois, a town of about forty-four thousand residents. DeKalb's early growth was sparked by the 1853 arrival of the Chicago and North Western Transportation Company rail line, and it was founded as a village in 1856. The village was incorporated into a city in 1877, four years after another technology drastically altered the identity of DeKalb and of the nation as a whole.[1]

It all began at the 1873 DeKalb County Fair. Henry Martin Rose of Waterman Station, Illinois, displayed a device to prevent stock from pushing against and through wire fences. Rose's "new and useful improvement" was a one-inch-square board with "metallic points . . . six or eight, more or less, inches apart," meant to be secured to preexisting smooth-wire fences by wire bands.[2] Although Rose's patent was a cumbersome solution to the fencing problem, his invention inspired three DeKalb residents who examined the display. Joseph

Figure 3. Portrait of Joseph Glidden. From the *DeKalb County Manufacturer*, 1882.

Glidden, a farmer; Jacob Haish, a lumberman and builder; and Isaac Ellwood, a hardware-store owner, all rushed to file patents for their own "improvements in wire fencing." All three patents employed wire "spurs," "prickers," or "spikes" held fast by two twisted wires. Glidden filed first. Haish challenged the filing and delayed approval of Glidden's patent until after his own patent was awarded. Ellwood set aside his patent and joined Glidden to produce and sell the "Glidden Coil." In 1876 Glidden sold his patent rights to the Massachusetts wire manufacturer Washburn & Moen, and Ellwood soon entered into a partnership with the company. The partners bought up competing patents for various forms of "thorny fences," creating a powerful monopoly.[3] Haish rebuffed their attempts to add his patent to their collection.

In 1882 the *DeKalb County Manufacturer* published a special edition that included an entire section on barbed wire.[4] The booklet contained numerous illustrations, including the portraits of the three men who had made a sleepy agricultural town into a bustling manufacturing center and who had become, as the portraits attest, prosperous captains of industry (figs. 3, 4, and 5). The publication featured lengthy statements by Isaac Ellwood and Jacob Haish. The facts they chose to present give readers a condensed history of the young barbed wire industry and insight into the personalities of the men whose ingenious products met a pent-up global demand for cheap and efficient fencing.

I. L. ELLWOOD,

Figure 4. Portrait of Isaac Ellwood. From the *DeKalb County Manufacturer*, 1882.

JACOB HAISH,

Figure 5. Portrait of Jacob Haish. From the *DeKalb County Manufacturer*, 1882.

Ellwood's account offers a history of fencing from Josephus's attribution of Cain as the first fence builder to the English Enclosure Acts. Citing Ohio, Kentucky, and Iowa agricultural reports that lament the cost of wood fences, Ellwood notes that in the United States there has been a long search for something "less expensive, more durable and effective" than traditional wood and stone fences. A history of smooth wire fencing concludes that although smooth wire "had no terrors for cattle," "a new era was dawning . . . the acme of fencing in all ages."[5]

Two illustrations from Ellwood's account demonstrate the progress from the old method of fencing to the new one (figs. 6 and 7). The tidy fields enclosed by strands of barbed wire are models of efficient control compared to the failing wood fences of the old system. The miserable log house has given way to the much larger house of a prosperous farmer. The out-of-control horse pulling the concerned driver of the carriage in the first cut is replaced by two horses well in hand. The new technology, the image asserts, enhances the pastoral setting. And to whom do readers owe thanks for the new fence that delivers order, control, and prosperity? To a genius:

> The honor of producing this improvement is due to J. F. Glidden, of DeKalb, who in May, 1874, was granted a patent on barbed wire, the

THE OLD SYSTEM OF FENCING.

Figure 6. The old system of fencing. From the *DeKalb County Manufacturer*, 1882.

THE NEW SYSTEM OF FENCING.

Figure 7. The new system of fencing. From the *DeKalb County Manufacturer*, 1882.

> essential feature of which was "coiling a short piece of wire between its ends around the fence wire." . . . That single idea has revolutionized the fence building of the world, and has accomplished what two generations past sought in vain to accomplish—universal availability, cheapness, durability and effectiveness in a fence. . . . It must take rank as one of the most beneficial products of American genius.[6]

As proof of the demand met by the product of Glidden's genius, Ellwood lists a remarkable set of production and sales figures:

> 1874 there were 10,000 lb made and sold
> 1875 there were 600,000 lb made and sold
> 1876 there were 2,840,000 lb made and sold
> 1877 there were 12,863,000 lb made and sold
> 1878 there were 26,655,000 lb made and sold
> 1879 there were 50,337,000 lb made and sold
> 1880 there were 80,500,000 lb made and sold
> 1881 there were 120,000,000 lb made and sold
> 1882 the estimate is for 180,000,000 lb[7]

Ellwood claims that cattle will find his "hedge of steel thorns" terrifying and simultaneously downplays complaints about the new fence: "Countless objections were urged against its use, the most prominent of which was, that it was a cruel, dangerous fence to turn stock against, a prejudice which is still entertained by a few old fogies in almost every community."[8] He continues his story with a description of the modern facilities for manufacturing barbed wire in his DeKalb factory (fig 8).[9] The engraving shows carloads of barbed wire exiting the DeKalb factory as the smoke of industry rises from the freight and passenger trains and from the factory itself. The enterprise is glorious proof of American progress through technology.

Following Ellwood's account, Jacob Haish describes his own heroic attempts to develop an economical and effective fence. He invokes ancient Greece and the Hebrew Bible and, oddly enough in that mytho-historical context, logical reasoning:

> With spartan heroism, Mr. Haish kept on his way . . . for with the vision of a prophet he looked down the vista of time and saw revealed unto him the mid-day glory and triumph of the fair and shapely form of

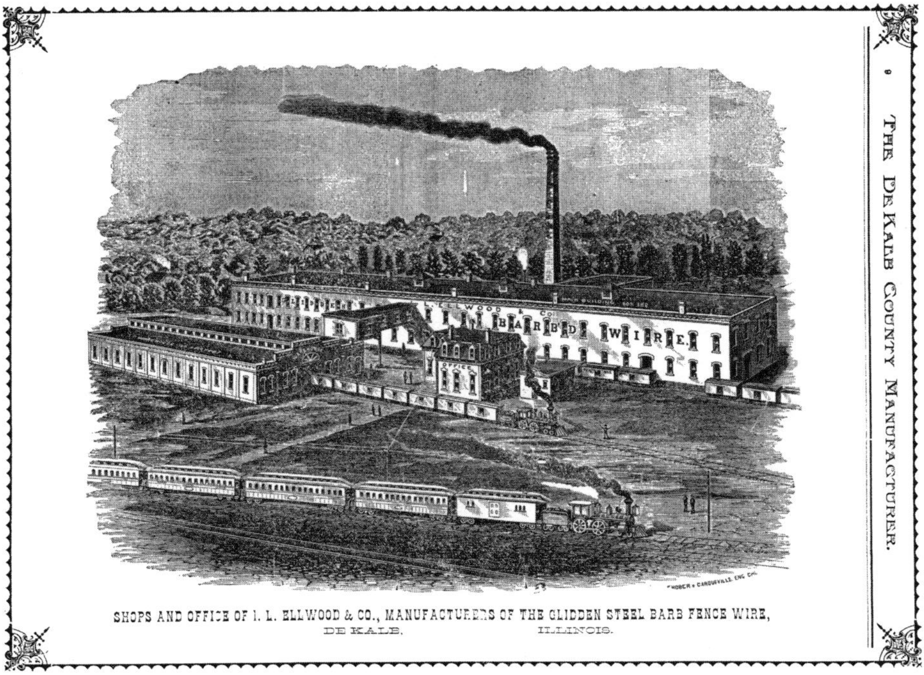

Figure 8. The shops and office of I. L. Ellwood & Co. in DeKalb, Illinois. From the *DeKalb County Manufacturer*, 1882.

> the "S" Barb, which was fast taking shape and comliness in the evolutions of the mind—the legitimate result of hours and days of anxious thought, and the result was the logical outcome of purely logical reasoning.[10]

"Truthfully," Haish concludes, he himself "is the father of the entire barb fence industry in the country. The leader in all improvements and patents, warranted to bring the business to its present magnificent growth." "The summing up of the matter," he continues, is that "Mr. Haish introduced one of the first successful barb wires, he made the first wooden spool, upon which to coil the wire for shipment; he used the first paint or varnish; he shipped the first spool of wire, via rail and water; he introduced the first automatic machinery for manufacturing barb wire."[11]

Ellwood and Haish thus present themselves to the public in the *Manufacturer*'s history of the already storied industry: the one man a genius, the other a prophet. As they became wealthy men and as their little town doubled and redoubled in size, Ellwood and Haish built palatial houses not far from each other, and Glidden built a hotel on DeKalb's main street to complete the intimate triangle. Although the *Manufacturer* accounts are mutually respectful (mostly by virtue of ignoring each other), personal encounters over the previous years had been tense, especially as Ellwood and his partners Washburn & Moen took Haish to court for patent infringement.

The antagonism between Haish and Ellwood was not one-sided, although some accounts describe Haish as the troublemaker and Ellwood and Washburn & Moen as the even-handed parties.[12] A letter dated November 6, 1876, and addressed to H. B. Sanborn, Ellwood's chief salesman (who married Glidden's niece and named his son Ellwood), reveals Ellwood's own hostility. That day, Ellwood reports, he signed and forwarded to his lawyers "the complaint against Haish—*it is a big one*, and carefully drawn, and the amt of damages and injunction bonds will make the Dutchman squeal."[13] The antipathy began early, and it was personal.

After the court decision in December 1880, there remained negotiations to be concluded. The Washburn & Moen Manufacturing Company owned the primary patents; that had been decided. But Haish owned the patent for a machine that both Ellwood and Washburn & Moen were using to manufacture their barbed wire. A series of as-yet-unpublished letters from Isaac Ellwood to his business partners in Massachusetts reveal his side of the tense interactions between the small-town neighbors as they struggle to come to terms.

> April 30th, 1881
>
> Gentlemen, . . . concerning the Haish matter &c., I am of the opinion that it would be useless to make Mr. Haish any offer for his patents at present, because the time is so short since the issuing of the last decree, that he does not realize the effect of it. . . . In regard to his actions, health &c., since this last decree, I have heard but very little, but what I have heard, it strikes me, that it has effected him more, and that he is feeling worse, and bluer than he has ever felt since the commencement of this litigation. I am told, that he is very poorly, and has kept in the house nearly all the time since the decree.[14]

Depression is understandable in light of the legal decision against Haish. Ellwood intimates that that could be good for the negotiations. Haish refused to comply. On June 8, 1881, Ellwood wrote that he had delivered a Washburn & Moen telegram to Haish and that Haish had "returned it enclosed in an envelope with one of his cards, which has the picture on it, of the spool rocking, wherein he states that he gets along just as well with no license, and without any comments whatever."[15] Negotiations continued, and Haish seemed close to an agreement.

> Mr. Haish and I met on the street, and after a considerable conversation on different subjects, Mr. H. managed indirectly to broach the present difficulty. After talking a few moments, I suggested to him that the street was hardly a proper place for us to talk over this matter, and told him that I would meet him at his office at some future time, when it would be most convenient for him. This was about noon, and he suggested that we meet at four, p.m. We did meet promptly on time, and had the usual skirmish lasting about three and one half hours. . . . I told him that I thought an arrangement could be made, that he should have four thousand tons free, three thousand tons additional at fifty cents per hundred pounds, and ten thousand dollars in cash. His last remark was, "make it five thousand tons free, and I will do it."[16]

When that possibility slipped away, Ellwood could not contain himself: "A word more with reference to the Haish matter. I am really of the opinion, that for inconsistency and unreasonableness, his equal is not to be found upon this green earth."[17] The negotiations that continued included the silent treatment, as reported vividly by Ellwood in late June:

> I am certainly through running after Mr. Haish. . . . A word more to show you a little how things work with him. While in Chicago, as usual he was taken with that peculiar disease of his, that seems to come at his will, at any time when he wished to postpone or delay any business matter. Mr. Haish and myself rode home on the same train, and he is a man who never uses any tobacco. That is he claims he can not stand tobacco smoke, and it is a very rare thing for me to find him in the car that I generally ride, which is the smoking car, but in coming home from Chicago Mr. H. took the smoking car all of the way, and in addition to this, he managed to change his seat on account of the smoke, and took one directly in front of me. . . . I continued to puff my Havanna [*sic*] as freely as if he was in a pullman. Of course his object

> was to get near me, thin[k]ing that I would say something in regard to the interview we had just had. But I never mentioned the subject, and I think he was somewhat disappointed in this respect. I simply mention these small details to you, in order to give an idea of the workings of this man Haish.[18]

Haish gave as good as he got. He published a story called "Smoke" in his almanac, the *Barb Fence Regulator*, that interprets the incident quite differently:

> An amusing incident occurred on the cars a short time since that will bear publishing. On board the train was a barb fence man, with a tired, unhappy look upon the otherwise smooth countenance, apparently seeking relief behind a choice Havana. At first the smoke wreaths almost enveloped him, but ere long the cigar falls from his hand, and he is in dreamland. The burdens of his dream came forth in murmurings low, yet distinct: "On the ragged edge of time I sit, / Bewailing my lack of sense: / To think that a weakling such as I / Could defeat the 'S' Barb Fence." Just here, the untimely whistle of the locomotive broke the sleeper's thoughts, and with a start he sprang to his feet, exclaiming . . . "where am I!" Moral—Don't talk in your sleep.[19]

A 1913 history of barbed wire written for the American Steel & Wire Company quotes a letter from Haish's lawyer dated June 29, 1881, the day before Haish finally proposed a settlement acceptable to Ellwood and Washburn & Moen. The letter urges Haish to lay aside a matter the lawyer finds trivial to get him "out of the clutches of the court, and out from under a heavy burden of possible damages." The lawyer ends his reasoning with an admonition that corroborates Ellwood's assessment of Haish's mercurial character: "Haish, for heavens sake, do think of this once deliberately before you make up your mind finally."[20]

The license that allowed Haish to continue manufacturing his "S" Barb with some restriction on tonnage and that gave Washburn & Moen the right to use Haish's machine for their production was finally granted on July 26, 1881. Soon thereafter Ellwood told his business partners that the agreement had been a wise choice:

> I really believe that, had no settlement been made, he would have caused us a great deal more trouble than we had anticipated. He

> showed me several cuts that he had made, when he supposed that our settlement would not be accomplished. In those cuts, you may imagine some of the most rediculous [*sic*] scenes, in regard to licensees paying royalties, as to monopolies, the indignation of the people, etc. etc.[21]

In *A Reminiscent Chapter from the Unwritten History of Barb Wire*, Haish notes that as the target of the litigation by Washburn & Moen, he had come to know "the greed and avarice that often impelled officers and stockholders of great corporations to pounce upon the weaker members of enterprises working along similar lines."[22] Haish points out somewhat gleefully that in the end Washburn & Moen spent 1.325 million dollars on the litigation, which resulted finally in the July settlement: "So I paid the Washburn & Moen Mfg. Co. 75 cents per 100 pounds royalty on wire and they paid me 75 cents per 100 pounds royalty on machines. Easy, wasn't it?"[23]

Although Haish and Ellwood had come to a fundamental agreement, aspects of the litigation dragged on until the US Supreme Court settled a dispute over the validity of various patents in 1892, finally declaring Glidden's "The Winner" the winner, although in surprisingly ambiguous (and somewhat awkward) terms:

> We are not satisfied, however, that he [Glidden] was not the originator of the combination claimed by him of the coiled barbed, locked and held in place by the intertwisted wire. . . . It was Glidden beyond question who first published this device, put it upon record, made use of it for a practical purpose, and gave it to the public. . . . Under these circumstances, we think the doubts we entertain concerning the actual inventor of this device should be resolved in favor of the patentee [Glidden].[24]

That an important legal case with national implications would play out between DeKalb neighbors Haish and Ellwood is as remarkable as the fact that tiny DeKalb itself was the initial center of a new industry that changed the face of America and the rest of the world over the next decades. It is a most intimate national story.

In the twenty-first century, the visitor who drives west from Chicago will find a town still profoundly marked by the invention and production of barbed wire. The toll road exit leads onto Annie Glidden Road. DeKalb boosters still call their town "Barb City," and the high-school mascot is "The Barb." The visitor will take note of the Glidden Crossroads shopping center, Glidden Campus Floral, the Barb City Manor, Barbed City Upholstery, the Haish Memorial Library, and the Hyrum Ellwood House. There is a Glidden Street, a Haish Street, and an Ellwood Street. The Ellwood House Museum, with its beautiful grounds, the sumptuous house itself, and an adjoining barbed wire museum, has become a cultural center for the town. Haish's ornate half-timbered house was torn down to make way for a parking lot for the Lutheran church. The town cemetery boasts elaborate monuments to Ellwood, Haish, and Glidden. Northern Illinois University (originally Northern Illinois State Normal School, now with more than twenty thousand students) was founded in part by large endowments from all three of the barbed wire inventors and manufacturers.

The history of the invention, manufacture, and use of barbed wire is fascinating. Even more interesting to us is the question of what barbed wire came to mean and the ways those meanings were constructed. In part 1, we examine several early constructions. One set of meanings was formulated through general and agrarian newspapers and magazines. Beginning with calls for effective and affordable fencing, print media covered the need, the discovery, the uses, and then the problems that became associated with the use of barbed wire. Another set of meanings came out of government inquiries into the alleged vicious nature of the fence, investigations instigated by the public clamor documented and fed by print media. These inquiries forced barbed wire manufacturers into a public defense of their product as state legislatures considered restrictive or anti–barbed wire laws. The public outcry also led to a flood of new patents aimed at mitigating the outrage. And finally, a third set of meanings was constructed through innovative and lively advertisements targeting the desires and fears of the American public.

Barbed wire is an object, a thing. Barbed wire is also a commodity, and anthropologist Arjun Appadurai argues that "commodities, like persons, have social lives." If they have social lives, he reasons, then the value of commodities is a politically mediated process, "consumption is subject to social control and political redefinition," and the "politics of value" is also "a politics of knowledge."[25] What values, we ask in this theoretical context, did barbed wire embody at the end of the nineteenth century? What social and political forces guided the consumption of barbed wire? What, in short, did barbed wire mean? Political scientist Jane Bennett takes the idea of the life of things one step further in her *Vibrant Matter: A Political Ecology of Things*. Attention to vibrant materiality, she writes, to the vitality of matter, "can chasten my fantasies of human mastery, highlight the common materiality of all that is, expose a wider distribution of agency, and reshape the self and its interests."[26] Our argument will be that barbed wire has in fact shaped our selves and our interests and that close attention to the steel fence as vibrant matter can disclose aspects of our shared past and help us reshape both selves and interests.

Our study of barbed wire is also informed by ideas related to "material culture," a term defined by historian Thomas Schlereth as "the study through artifacts (and other pertinent historical evidence) of the belief systems—the values, ideas, attitudes, and assumptions—of a particular community or society, usually across time."[27] Barbed wire belongs to what anthropologist James Deetz describes as "the vast universe of objects used by humankind to cope with the physical world." These artifacts are "the product of our thoughts . . . that sector of our physical environment that we modify through culturally determined behavior."[28] The object that is the subject of this book is an artifact made up of steel wires twisted around sharp barbs, a thing that has embodied and evoked ideas still important to us today. In the following three chapters we investigate that wire and those ideas as developed and contested in newspapers, as argued before legislatures, as modified in various new patents, and as sold through advertising.

Whether a thing or a commodity, whether vibrant matter or artifact, any object with a social life lends itself (and we might say avidly lends itself) to study. The result has been a most interesting body of scholarship about various objects. Sterling Evans, for instance, focuses on the unlikely subject of binding twine in his multifaceted *Bound in Twine*. Following what he calls the "henequen-wheat complex" from the Yucatán Peninsula of Mexico through the grain-growing regions of the United States and Canada, Evans produces a "case study of the social and cultural biography-of-things model recently advanced by anthropologists."[29] Transnational dependence on wheat crops and henequen harvests, Evans reveals, leads to social and environmental transformations, imported slave labor, gunboat diplomacy, prison-made twine, and an eventual crash when the newly invented combine made binder twine superfluous. Similarly, Gregory Cushman's *Guano and the Opening of the Pacific World* is a global ecological history focused on a single commodity. Cushman's obsession with bird poop leads him to the bird conservation program meant to maximize fertilizer production in Peru, to Aldo Leopold's influential land ethic, to the economic rise of New Zealand and Australia, and to issues of identity in several Pacific peoples. "This book demonstrates,"

Cushman writes, "that marine bird excrement is at the root of modern existence and fundamental to the incorporation of the Pacific Ocean into global history."[30] Those are striking conclusions rising out of an unlikely and smelly source. Witold Rybczynski, architect and professor of urban studies at University of Pennsylvania, was commissioned in 1999 by the *New York Times* to write an essay about the most useful tool of the millennium. The result was the book *One Good Turn: A Natural History of the Screwdriver and the Screw.*[31] Searching for origins of the mundane tool, Rybczynski's research moved from the screwdriver to the threads of the screw, helix-formed threads based on the work of mathematicians in Hellenistic Greece. He concludes that the helix was one of the most important principles of the Industrial Revolution and the mass production of interchangeable parts.

There are many examples of this kind of study: works, for instance, on the pencil, on salt, and on the zipper. They all share an interdisciplinary approach and a focus on the interface between the human and the nonhuman, on the interactive relationships between the material object and the maker and consumer of the object, and in each case the tight focus results in surprising conclusions about human history and identity. Our work on barbed wire does so as well.

Because barbed wire played a major role in agrarian history and the settlement of the American West, there is a substantial body of secondary literature about the fence. That work falls into essentially four categories: (1) barbed wire collecting, (2) individual wire company histories, (3) the significance of barbed wire in the late nineteenth-century American West, and (4) barbed wire usage in wars and prisons in the twentieth and twenty-first centuries.

Interest in the identification and classification of barbed wire throughout the twentieth century resulted in the publication of numerous works for collectors. Jack Glover's *The "Bobbed" Wire Bible*, for instance, an "illustrated guide to identification and classification," was published in seven revised editions between 1967 and 1986 alone. As collectors of barbed wire grew in number, more scholarly reference materials such as R. J. Thurgood's *The Complete Encyclopedia of Barbed Wire* (1979 and 1993) were published. Harold Hagemeier's *Barbed Wire: Identification Encyclopedia* is now in its fifth edition (2010).[32]

Some of the earliest and most-cited barbed wire histories are the 1920s company studies written by Arthur G. Warren, curator of the American Steel & Wire Museum. Warren's accounts outline the history of what became the most important wire company in the United States. Warren looks closely at the Massachusetts-based wire company of Washburn & Moen, which purchased Joseph Glidden's half-interest in the Barb Fence Company to create a joint company with I. L. Ellwood. The American Steel & Wire Company was incorporated, finally, into J. Pierpont Morgan's massive United States Steel Corporation. Warren's corporate histories are heavily rooted in the patent history of barbed wire and the litigation between barbed wire manufacturers, with only a slight nod to the social and historical consequences of this invention. A more thorough examination of the meaning of barbed wire in US history would have to wait another decade.[33]

One of the most significant works on the American West, Walter Prescott Webb's *The Great Plains*, includes a chapter on transportation and fencing that begins to touch on the effects of barbed wire on "human life in the Great Plains." Webb argues that "the Industrial Revolution found itself confronted by four major problems in the Plains country: (1) transportation, (2) fencing, (3) water, and (4) farming. The first was solved by railroads, the second by barbed wire." Webb remarks that "the story of the effects of barbed wire on human life in the Great Plains is one that has not been and cannot be adequately told." Along with its revolutionary effect on ranching, barbed wire fencing also "revolutionized land values and opened up to the homesteader the fertile Prairie Plains,"[34] not to mention contributing to the conquest of Native Americans in the American West.

A professor at DeKalb's Northern Illinois State Normal School, Earl Hayter, did the best early scholarly work on the history of barbed wire. His 1939 article "Barbed Wire Fencing—a Prairie Invention" investigates how barbed wire radically altered the social, economic, and political direction of western development. After the Homestead Act of 1862 and the first sales of barbed wire in 1874–1875, small farmers used barbed wire to settle the Plains. Their movement prompted cattlemen first to defend free-range lands and then to secure the land by fencing it. What resulted was not a peaceful carving up of the public domain but a series of disputes that pitted farmer against rancher, big ranchers against little ranchers, sheepmen against cattlemen, fence advocates against fence cutters, and eventually the rise of stock grower associations and protective agencies that protested the barbed wire monopolies. One by one, Hayter describes how these disputes were largely resolved within two decades of barbed wire's appearance on the Great Plains.

With the use of barbed wire, Hayter writes, farmers produced better crops and managed their lands more effectively. Cattlemen also produced a higher grade of beef by confining cattle in fields to feed on range grasses, hay, and eventually corn and alfalfa grown in fields protected by barbed wire. Hayter also points out that barbed wire left its political stamp on the West in the lobbying efforts by proponents and opponents of barbed wire, the hotly contested political debates surrounding the enclosure of the public domain, and the long-running patent wars among barbed wire manufacturers. By the end of the nineteenth century, the West dominated by cattle companies and "cow culture" was transformed through the use of barbed wire into a new era of small farmers and stockmen.[35]

One of the topics Hayter barely touched on in 1939, the fencing of railroad rights-of-way, became the basis for his 1945 article "The Fencing of Western Railways." In this study, Hayter interweaves two key technologies that opened and settled the American West, specifically focusing on the introduction of the railroad into the West and how it came into conflict with the dominant open-range cattle industry. His discussion of the eastern common law practice of enclosure, which required owners to fence in their animals or be liable for trespass and damages, and western states' herd laws, which permitted cattle to run freely, provides insight into why western law soon required railways to fence in their tracks. From the earliest application of barbed wire fencing on the Chicago and Northwestern Line in 1877 (the line that ran through DeKalb), Hayter describes a spate of lawsuits brought against railroads for the livestock injuries sustained, ironically, as a result of fencing. Although barbed wire's reputation as a cruel fence would decline over time, it was the introduction of the woven wire fence that allowed most western railroads to fence their rights-of-way with less controversy by the end of the nineteenth century.[36]

Another topic raised by both Webb and Hayter, but not pursued, was the issue of fence cutting. Wayne Gard's 1947 article "The Fence-Cutters" examines the impact of barbed wire fencing in Texas and the backlash it initially raised. In particular, he examines various motives behind fence cutting, whether it came from farmers denied access to water, small ranchers denied access to grazing, or landless cowboys now unemployed as barbed wire fences took over some of their functions. Gard concludes that despite the "inequities and temporary hardships" barbed wire wrought in Texas, the presence of barbed wire in the West signaled the era of permanent settlement.[37]

C. Boone McClure's brief *History of the Manufacture of Barbed Wire* published in 1958 features an analysis of the vitriolic litigation between barbed wire manufacturers over patents and licenses and provides a sense for the decades of conflict that took place in the boardrooms and courtrooms of the East.[38] Joseph McFadden's 1978 article about the competition between barbed wire manufacturers and their desire to create a patent monopoly in the fencing industry in the years 1873–1880 details how Ellwood and Washburn & Moen campaigned against competitors to purchase their patents or scare them out of the business through threatened and actual litigation. By 1879, the industrialists had achieved some success in forcing small competitors to fall in line, but their biggest competitors refused to submit. The renegade competitors began to file federal lawsuits, which ultimately failed. The 1880 court decision resulted in the closure of many barbed wire manufacturing plants as well as the restriction of production in factories that continued to produce barbed wire.[39]

A comprehensive history of barbed wire in the American West finally appeared in 1965.[40] Henry and Frances McCallum's *The Wire That Fenced the West* is still the standard synthesis of scholarly works on barbed wire and thus merits close attention. The authors rely heavily on wire companies' records, agricultural societies' reports, newspaper accounts, and private papers of barbed wire competitors, as well as the earlier works by Warren, Webb, and Hayter.

The book is divided into three parts: "Barbed-Wire Fence-Makers," "Barbed-Wire Fence-Builders," and "Types of Barbed Wire." The first section begins with one explanation of why the Great Plains was such a bulwark to settlement: the lack of inexpensive and suitable fencing. Once wire-making methods improved with the use of Bessemer furnaces and Bedson rolling processes, wire manufacturing experienced rapid growth. Smooth wire, however, was ineffective in keeping livestock and humans in or out of fields. Something else was required. That something else, barbs attached to wire, first surfaced in the late 1860s and was perfected throughout the 1870s and 1880s. The McCallums detail the prehistory and history of barbed wire, its inventors, and promoters. These are lively, interesting accounts like the iconic story, for instance, of John "Bet-a-Million" Gates, who was sent to San Antonio, Texas, at the age of twenty-one to sell barbed wire. He had little success until he built a barbed wire corral in a plaza and stocked it with Longhorn steers. While a curious crowd watched, the steers tried to run through what was for them an invisible fence. Repulsed, the enraged steers charged again and again until they were finally subdued. Barbed wire sales boomed in the Southwest, and Gates's fortune was secured.[41] And finally, the authors discuss the cutthroat competition among barbed wire manufacturers, the patent litigation that resulted from this competition, and how Ellwood and Washburn & Moen established a near monopoly on barbed wire production.

The second part of the McCallums' book covers the transference of barbed wire to the Plains, with particular emphasis on Texas as cattlemen and farmers vied for control of the open range. Noteworthy is the authors' discussion of livestock injuries and deaths due to physical encounters with barbed wire fencing. Cuts, screwworms, and maimings were problems, and the large numbers of cattle deaths in the winters of the early and mid-1880s resulting from barbed wire drift fences across the Plains were shocking. The McCallums' research on several big cattlemen and cattle outfits provides insight into the shift they made in regard to barbed wire: from opposing the use of barbed wire by settlers to the eventual use of barbed wire to expand and protect their private empires. The authors conclude, as did Webb and Hayter, that the introduction of barbed wire in the Great Plains changed the course of development in this region. The firstcomers, Abel Pierce, Charles Goodnight, and John Chisum, were gradually and sometimes violently replaced by railroads, sheepmen, small ranchers, and homesteading farmers. Barbed wire was the death knell to the open range.

The last section of *The Wire That Fenced the West* reviews different barbed wire classifications (Early Varied, Vicious, Obvious, Modified) and identification of key barbed wire types. There is also a brief discussion of twentieth-century uses of barbed wire in wars and prisons.

By virtue of its scope, perhaps, or because other perspectives were not forthcoming, the McCallums' book remained the last word on the subject for nearly forty years. Early in the twenty-first century that state of affairs shifted radically. In quick succession, three new books took up political and cultural questions raised by barbed wire. In *Barbed Wire: An Ecology of Modernity*, Reviel Netz's analysis of the ecologies established by deployment of barbed wire fences emphasizes the violence required for effective control: "as iron (and, most important, steel) became increasingly inexpensive and widespread, it was used to control motion and space, on a massive scale, exploiting its capacity for mass production and its power of violence over flesh."[42] Netz argues that the invention and manufacture of barbed wire for the voracious agricultural market made possible its use in the trenches of World War I and in the concentration camps of the Boer War and of World War II. All these events, Netz writes, "involved, on a mass scale, control over space, which is tantamount to the prevention of motion, which is tantamount to violence."[43]

Olivier Razac's short *Barbed Wire: A Political History* emphasizes what he sees as the depredation of Native American nomadic culture at the hands of the stringers of barbed wire. The new fence, he argues, closed off Indian lands so "the white man could occupy and exploit them," and it "chopped space into little bits and broke up the communal structure of Indian society. Barbed wire made the Indians' geographical and social environment hostile to them." Razac writes that even as its use has waned with the rise of alternative tools of surveillance and constraint, barbed wire remains a potent symbol of "the various barbarities and tyrannies with which it is associated."[44]

A third book, Alan Krell's *The Devil's Rope: A Cultural History of Barbed Wire*, examines what barbed wire has come to mean, as opposed to its role in history. Krell analyzes a vast array of images—propaganda leaflets, photographs, paintings, works of installation art, films, and poetry—to uncover a complicated nexus of constraint and control in the various representations of barbed wire. He addresses "those nuanced, figurative spaces that barbed wire can occupy: between threat and security, culture and nature, the familiar and the unknown."[45] Although our focus is more historical and literary than Krell's, this complex of meanings is at the heart of our book as well.

While an extensive barbed wire historiography exists, little scholarly attention is given to the basic question we ask in part I: How did the entities involved in the invention, manufacture, marketing, and use of barbed wire deal with the contradictory nature of this invention? Barbed wire is efficient in part because it is dangerous. Advocates of barbed wire emphasized the efficiency. Opponents highlighted the danger. As they did so, all parties helped construct the meanings of a radical new fence, meanings that make up the collective identity of barbed wire.

CHAPTER ONE

"Infernal Machines"

Debating the Meaning of Barbed Wire Fences in the Media

[Barbed wire] is capable of resisting the assaults of the most vicious or unruly animals. Its sharp pointed barbs act as spurs to inflict pain upon the sentient nerves of the animal, which lie close to the surface of the skin, and deter it, without injuring the animal, from pressing against the fence.

—B. F. Thurston, chief counsel for the Washburn & Moen Manufacturing Company, 1880

Wire fences are death dealing, infernal machines and should be wiped from the face of the earth.

—*Saline County Journal*, 1883

Barbed wire controls animals by inflicting pain. That makes it a cruel fence. An early protest against this particular form of cruelty occurred in 1857 in Austin, Texas, after a Swiss American iron maker named John Grenniger attached jagged metal strips to the top of a fence around his orchard. Neighbors were outraged at the damage done to their marauding stock and spent years trying to run Grenniger out of town, efforts that ended only after his murder in 1862.[1] Inventors of such early "improvised" technologies were, as the McCallums point out, "hesitant or unwilling to own up to the discovery they had made, mainly because it was not in keeping with the sentiment of their time. To present sharp projections to livestock . . . was intolerable."[2]

The demand for effective fencing, however, and the proven effectiveness of barbed wire, meant that the new barbed wire fences rapidly crisscrossed the country. Barbed wire was quickly associated with agricultural development, westward expansion, and societal progress. The celebrated new technology also developed a reputation as a vicious implement responsible for injury and death. Both positive and negative associations were developed, in part, through a vigorous public debate that played out across the nation. In this chapter we survey dozens of the newspapers and magazines in which the debate took place, uncovering surprising stories, rhetorical bombast, and nuanced arguments. Many accounts were simply factual. They announced the arrival of barbed wire, described how it differed in design from previous fencing, and documented the appearance of agents and factories across the nation. Barbed wire was praised as ornamental and as visible proof of good husbandry. There were stories about usage along railroad rights-of-way and in urban settings as public safety and trespass solutions. Some accounts raised the specter of monopoly. And there was considerable outrage over the "barbarous" fence. Eventually, as one might expect in the context of such widespread interest, barbed wire took on meanings beyond the contested claims of the public debate.

The Problem and the Answer: The Perfect Fence

"Crowds of farmers visited the Grange Store yesterday to inspect the barbed wire fence."[3] According to the *Los Angeles Daily Herald*, the advent of barbed wire was big news in Los Angeles in 1875. That same year, the *Wichita City Eagle* aptly summed up the nation's fencing problem and what it saw as the solution:

> The fences of the United States are reported to be worth \$1,800,000,000, or \$45 per head of each inhabitant. . . . The expense necessary to keep them good, is at least \$100,000,000. Major Brisbane, of the United States Army, estimates the annual destruction of timber in the United States at 5,500,000 acres, and that one-half this immense area is required for fencing alone. However true this may be, it is nevertheless true that fencing is one of the most expensive items the farmer has to deal with. The increasing scarcity of timber in the West has caused the total abolishment of fences over large areas. . . . Barbed Wire in its various forms for fencing, now constitutes an immense and constantly increasing business.[4]

In 1877, the *Cultivator & Country Gentleman*, after commenting on the trouble with board fences, saw barbed wire as a good alternative: "The best remedy we have yet seen for this trouble is the new barbed wire. . . . Horses and colts which have been in the practice of leaning against fences and crowding off the boards, or leaping over them . . . maintain a respectful distance."[5] Two years later, the *American Agriculturalist* addressed the fencing problem "on the treeless prairie" between the Mississippi River and the Rocky Mountains:

> For a long time simple wires were resorted to and the chief efforts at improvement looked to devices for keeping up the proper tension during the expansion and contraction of hot and cold weather. But they have always been unsatisfactory. The animals would run against and break them. They would crowd their heads, and often their bodies, between the wires in reaching for a choice bit of better grass, or perchance a cultivated crop, on the other side. Within a few years, various arrangements of Barbs, or Sharp Points, at intervals along the wires, have been provided, and double or triple wire used, twisted together like the strands of a rope, both to hold the barbs and to diminish the difficulty arising from expansion and contraction of the wires. These were considered a great advance, and have largely been adopted.[6]

In these early years, many newspapers ran stories praising the new fence in language that echoed manufacturers' claims that this was "the perfect fence."[7] The 1875 *Los Angeles Daily Herald* piece cited previously, for instance, proclaimed that "the unanimous verdict was that it was the best and cheapest fencing in the world. . . . It takes less posts . . . less labor; is the most durable. . . . It costs one-third less than the commonest kind of poor fence, and for this reason alone particularly recommends itself to farmers."[8] A week later the *Daily Herald* repeated and amplified its praise, claiming that farmers had "pronounced it the fence of the period. With cheapness, it combines durability, making it the most desirable fence ever presented to the public."[9] In December 1875, under a regular column titled "Local Intelligence," the *Waco Daily Examiner* recommended "the barbed fence wire, patented by J. F. Glidden, which is strongly endorsed by Northern farmers, who use it very extensively. Our farmer friends should see this great improvement in fencing wire."[10] In an 1877 story about new barbed wire near Emporia, Kansas, the *Emporia News* noted, "Some of the farmers are beginning to talk those [barbed wire fences] up as the best fence for the money, in this locality."[11] Pennsylvania's *Elk County Advocate* wrote in 1878 that barbed fencing had the "highest favor with landowners, for farm and home protection against depredators of the crops and orchards."[12] In a December 1882 letter to the editors of the *Cultivator &*

Country Gentleman titled “The Fence of the Future,” J. W. Clarke of Plymouth County, Iowa, claimed, “Barbed wire is now known to be the most effective fence in use; and its use is very rapidly extending in the West, and considerably in the Middle States and eastward. Barbed wire fence will become the fence of the future.”[13] In 1883, the *Butler Weekly Times* saw barbed wire as a boon to enterprising settlers: “Now-a-days with sawed lumber, barbed wire, and hedges, with coal for fuel, the enterprising settler can locate himself in the middle of the wide prairies with a supreme indifference to the whereabouts of the timber.”[14] Even somewhat reluctant commentators recognized the importance of the new fence, as noted in 1883 in the *Phillipsburg Herald*: “The barbed wire is certainly a great invention, and while we cannot say that we like it as a fencing material, there is no denying its value for all regions where timber is scarce or dear.”[15] The *Herald* did not elaborate further on why it did not like barbed wire.

Barbed Fencing: Technological Wonder

As barbed wire usage expanded from coast to coast, the media that catered to farmers and ranchers marveled at the technology of the new fencing. The *Elk County Advocate* printed a description so that readers not yet familiar with the fence could visualize it: “The Barb Fence is literally a steel thorn hedge with all the defensive qualities of the most perfect hedge, and all the durability of galvanized steel wire.”[16] Throughout the 1870s, newspapers and magazines found the appearance of the new technology in their locales newsworthy. In a statement that calls the fence “wire barbed,” like other sometimes awkward reports on a product still finding its way into public consciousness, the *Emporia News* noted in 1877 that “E. McMillan and J. K. Hewitt are putting up a wire barbed fence between them.”[17] The *Arizona Sentinel* reported in 1878 that “Geo. Mowry, of Phoenix, has some rolls of patent barbed wire for fencing; thorny looking stuff to handle. . . . These little things are what give one an idea of the rapid growth of the Territory.”[18] The *Cultivator & Country Gentleman* alerted its agrarian readership to the technology of barbed wire in a 1877 article on the New York State Fair’s Implements and Machinery display: “In the same building, the machine for making Washburn & Co.’s barbed wire fencing, from Worcester, Mass., was in operation. It takes the galvanized steel wire from the coils, twists it into a rope, at the same time cutting, forming and inserting the barbs. It made about 25 feet in a minute.”[19] The St. Paul, Minnesota, *Daily Globe* reported in its 1879 State Fair column that “I. L. Elwood & Co., manufacturers of Glidden's patent steel barb fence wire, DeKalb, Ills., through their agent, Mr. Stone, have secured space, and are now putting up shafting, belts, etc. to manufacture during each day of the fair barbed fence wire." The appearance of this new technology was an item of interest to more than just the agricultural community, it was in fact, as the *Globe* noted, "a leading attraction and will be worth going a long way to see."[20] This was echoed in the 1883 "Local Matters" column of the *Iola Register*: "A large crowd of interested spectators were watching the process of barb-wire making at the factory last Monday."[21] And less than a month later, the *Register* noted that "the barb wire factory is putting in full time now, turning out about 2,000 pounds daily. That the wire gives good satisfaction is shown by the fact that it is sold as fast as made."[22] Ten years after its introduction, barbed wire still maintained its novelty, as exemplified by *Century Magazine*'s coverage of the 1885 New Orleans Exposition: "The machines at work making barbed wire afford a glimpse of a new and important industry which has grown up with the settlement of the treeless plains of the Far West."[23]

Coverage of the arrival of barbed wire and the spectacle of manufacture was accompanied by stories of distribution. Visits by agents of the new industry and early mass shipments of barbed fencing to stores were newsworthy. In 1878 the *Rock Island Argus* included under its "Personal" column J. M. Montgomery's trip to Joliet to "investigate the hardware trade," where he "bought a car load of barbed wire fence from the Joliet Wire Fence Co., a concern that sells $500,000 every year."[24] A month later the *Argus* noted that "W. J. Adam, of the Adam Manufacturing Co., of Joliet, manufacturers of the steel barbed wire fence, paid a flying visit to Rock Island to-day."[25] The Dakota Territory's *Bismarck Tribune* noted a conjunction of manufacturer, agent, and user in a brief story in June 1879: "W. G. Stiles, of the Thorn Wire Hedge Co., Chicago, was in the city this week, interviewing his agent, Mr. Peoples, on barb wire, etc. He thinks the barb will come in general use in this section."[26] Announcements like these were common in print across the United States well into the 1880s.

Good Husbandry, Beauty, and the Eye of the Beholder

Some early accounts associated the new fencing with elements of good husbandry and with aesthetic properties. Articles of this sort appeared in sources as geographically diverse as the *Weekly Arizona Miner* and the *Middlebury Register*. From the *Miner*: "A large lot of barbed fence wire for C. Becker, of Kirkland Valley, has arrived, and that gentleman will soon have his fine farm enclosed with a good wire fence"; and from the *Register*: "Philip Stone has removed his dooryard fence and replaced it with a handsomely arranged barbed wire fence."[27] In 1879, the Morrisville, Vermont, *News and Citizen* noted both the efficacy and the aesthetics of some of its local residents' new fencing: "H. Morse and J. Green have put up a long string of barbed wire fence this spring, which proves to be an effectual barrier to cattle, besides being nice and neat looking."[28] *The Iola Register* reported in 1880 that "the U.B. [United Brethren] Church that has been under process of erection at this place for some time past is now completed. It is an elegant structure 28 × 42 feet, with a good inside finish, well seated, and is surrounded with a neat barbed wire fence, hitch racks, and other things that adds to the beauty and convenience of the location."[29] In 1882, the *Nebraska Advertiser* commented on the use of barbed wire in Nemaha City, Nebraska, noting, "City Hall Square is worthy of mention. Last year it was fenced with posts and barbed wire, trees planted and blue grass seed sown . . . the whole giving the square a very attractive appearance."[30] In the same year, the *Cultivator & Country Gentleman* included a letter from a reader in Lexington, Kentucky, that observed, "There is a park in this city fenced in this way [barbed wire strung along a hedge], and it is very ornamental as well as impervious to stock or boys."[31] The *Dallas Daily Herald*'s 1883 column "Texas Press Opinions: Dawn of a Better Day" was optimistic about the advent of barbed wire, citing the new fencing as a civilizing improvement to the rural scene. That the author was aware of the fence's potential for harm is clear from the fact that he describes it as only "slightly barbed":

> Among the many gratifying scenes to be witnessed by the traveler through this county is the improvement in house building. . . . The fencing, too, is beginning to show a marked improvement, the old time being supplanted by the more durable and slightly barbed wire fence. In fact, a visit in any direction will force upon an observer the marked degree of improvement going on in this and adjoining counties.[32]

A decade later, the *Southern Cultivator and Dixie Farmer* was still touting the utilitarian and ornamental nature of barbed fencing and using the familiar phrase "the fence of the future": "That the fencing of the future will be wire is now clearly indicated. . . . Its cheapness and durability are factors which will determine its general adoption, and the many handsome designs into which it is now worked are bringing it also into great use for ornamental fencing in cities and villages."[33]

Not all editors and reporters remarked positively on the aesthetic qualities of barbed fencing. In a story titled "Outside Echoes: Beautifying a Prairie Home," the *Press and Daily Dakotaian* touted the sylvan beauty of a residence with many trees and hedges over the less attractive barbed wire: "What a paradise judicious and continued silva culture can make a prairie home. . . . A beautiful hedge adorns the eastern boundary of the farm and in quite a contrast with the barbed wire fence in the locality."[34] With similar sensibilities, Frank French noted in a piece about country roads for *Scribner's Magazine* that "the stone walls of New England are eminently appropriate and picturesque. . . . So many elements of beauty, of interest, of utility and appropriateness dwell within them. . . . It may be easier to stretch a barbed wire above than to repair their gaps, but how degrading the effect!"[35] The same *Iola Register* that three years earlier had praised a church's "neat barbed wire fence" reported in 1883 that "the commissioners of Woodson County ordered a barb wire fence built around the square at Yates Center. This did not meet with the approval of the citizens, and a collection was taken up to build a pine board fence."[36] Under its "Neighborhood News" column, the *Wheeling Daily Intelligencer* reported on an unfortunate consequence of barbed fencing in urban/public spaces:

> That line of barbed wire stretched from the Court House gate to near the steps should be removed at once. Your correspondent saw two boys run against it within the space of ten minutes a few evenings ago, one of them having his throat badly lacerated. He was taken to Dr. Davis' drug store for treatment, and before he left the Doctor's his brother met with the same accident, but fortunately did not receive such serious injuries.[37]

Similarly, the St. Paul *Daily Globe* reprinted a story from the Le Sueur *Sentinel* that questioned the intelligence of allowing such a dangerous fence in public spaces:

> It does not strike the average looker on as just the best thing to have a barbed wire fence enclosing several lots in the heart of the town, and especially adjacent to the business center. There are several considerations, aside from the appearance of the thing, one of which is the danger to life and limb of persons. . . . Perhaps no other town in the state would think of erecting such a fence in such a location, much less tolerate it.[38]

Barbed fencing may not have meshed with all public spaces, but many newspapers and magazines initially reported to their readers that there was at least one public thoroughfare that benefited from its placement: the railroad right-of-way.

Railroad Rights-of-Way

In March 1878, the *Marshall County Republican* in Plymouth, Indiana, noted the arrival of barbed wire along a railroad right-of-way with a touch of humor: "The Illinois Central railroad are using barbed wire for fences along their line of road. Where will the poor tramp rest his weary aching limbs now."[39] Under its "Neighborhood News" column of November 11, 1880, the *Northern Pacific Farmer* applauded railroad usage of barbed wire in Minnesota: "The railroad company have a crew of men engaged in building fence along the track. They are using cedar posts and barbed wire, which is said to make a very efficient fence. This is a much needed work, as cattle killing was getting altogether too common."[40] The following month, Louisiana's *St. Landry Democrat* posted a notice for its readers from the general manager of the Louisiana Western Railroad:

> That whereas . . . the killing and maiming of stock by heavy trains . . . has been considerable and unavoidable . . . the General Manager . . . has determined to extend to the residents along the line of his road, the means for such protection, by furnishing sufficient barbed iron wire and fastenings for a fence of two wires on each side of the road bed, to all who will apply for the same and sign an obligation to put up (furnishing the posts and labor) and thereafter maintain a fence on each side of the road bed.[41]

In 1882 the Stanford, Kentucky, *Semi-Weekly Interior Journal* posted a story under its "Notes of Current Events" column that praised the economic advantages of barbed wire: "Barbed wire fences have been built along the entire line of the Nashville, Chattanooga and St. Louis railway. It has already paid for itself in protecting the road from cattle, which had to be paid for when killed. It also saves the road from destructive wrecks on account of trains running over cattle."[42] The *Milan Exchange* thought it still newsworthy in April 1883 to note the Chesapeake, Ohio & Southwestern Railroad's new contract "to build a barbed wire fence along both sides of the railroad from Rives station to Memphis, enclosing all the company's land. This is done to prevent killing stock. The fence will cost the company over a hundred thousand dollars."[43]

Even as the print media reported that barbed wire resolved the agrarian fencing need, the railroads' most pressing public relations problem, and even some public safety and security issues, the actual use of the new fencing material created problems for both animals and humans. Almost in tandem with heralding barbed wire as the invention of the age, newspapers and magazines began to report on the dangers posed by the "Perfect Fence."

A "Barbarous" Fence

An 1882 piece from Denver's *Daily News* laid out the problem in graphic terms:

> It would be a noble work if the [Humane Society] could prevail upon the Legislature to pass a law prohibiting the erection of "barbed wire" fences. . . . Scarcely a day passes but one can hear of the death or fatal injury of a cow, calf, horse or colt which has run into the fence (which cannot be seen far away), and so cuts its head and body as to result seriously and often fatally. . . . The foregoing is brought out by a sickening sight that met the eyes of passengers on the accommodation train between Collins and Loveland, yesterday morning. A herd of milk cows was feeding along the line of the railroad track, and as the train came rolling along a number of the cows started to cross the track. The engineer blew his whistle loud and shrill, and the frightened cows began to run in every direction. One unfortunate ran headlong into a wire fence near the track and jumped head and fore legs through the fence and there hung on the sharp barbs, and as the rear car passed by the passengers saw the brute hanging there on the sharp barbs kicking and bellowing, the piercing instruments sawing and cutting into her body deeper and deeper as she struggled for liberty. If the man who built the fence had a heart, not as hard as stone, and could have seen that terrible butchering, he would have solemnly sworn never again to build a barbed wire fence.[44]

As early as 1878, the *Leavenworth Weekly Times* included a page-one story about a passing train frightening a herd of horses that had run into the barbed wire fencing. The story concluded that "the place mentioned is a very dangerous one, and we have been expecting an accident of this kind for some time."[45]

What began as a trickle of animal injury stories became an avalanche. That same year the *Dodge City Times* offered a dire prediction: "Barbed fence wire is doing a great deal of damage to stock in Colorado and Kansas. The Citizens won't stand for it very long."[46] From the *Emporia News* on August 9:

> A good horse, worth $75, belonging to Lemuel Johnson, ran into a barbed wire fence on the 1st, and was so terribly cut that it died the same night. Two other of his horses have been badly cut in such fences, but have recovered. Mr. Johnson thinks it is time there was some kind of law relative to hedge fences.[47]

On September 5, 1878, two Kansas newspapers reported on separate events. In Troy: "W. A. Byers had a horse badly injured, by getting tangled in a barbed wire fence." And in Leavenworth:

> On last Saturday four gentlemen from Missouri were passing west, on the road east of Troy, with 22 horses which they were taking out west for sale. Near John Rhue's is a barbed wire fence, close to the Rail road track. While passing this place a train on the Denver Road came along, which frightened their horses, and they ran into the wire fence, wounding several,—two of them so badly they had to dispose of them, selling them both for eighty dollars. Several more were considerably hurt, but they were able to travel.[48]

By 1879, accounts of horses injured by barbed wire were common in newspapers and magazines across the nation. Two stories from the *Lexington Weekly Intelligencer* of that year reported, for instance, that "Mr. Waterhouse, near Mayview, had a fine horse ruined last Friday by running against such a fence" and that "last Saturday Mr. James McGirl . . . had two fine mules almost ruined by running against a barbed wire fence. . . . But a few weeks ago Mr. H. Waterhouse, near Mayview, had a fine horse nearly ruined in the same manner."[49] That same year the *Lewiston Teller* reported that "a horse belonging to Mr. Hutchings ran against the barbed wire fence of Judge Adams and become terribly lacerated."[50] In August, the St. Paul, Minnesota, *Daily Globe* noted under its "Neighborhood News" that "Wm. Coleman, a farmer living in Wheeling, had a fine colt very seriously injured Saturday by running against a barbed wire fence on his place."[51] Reprinted stories amplified the effect of anecdotal tragic encounters. An 1879 issue of the *New Orleans Daily Democrat*, for instance, reprinted a Charlotte *Observer* report that questioned the ability of the new fence to keep cows away from railroad tracks: "Seventy-eight cows were killed on the right of way of the Atlanta and Charlotte Air-line Railroad during the month of July, notwithstanding that many miles of barbed wire have been put up on either side of the track to keep cattle off."[52]

In September 1880, the *Emporia News* ran a brief note pointing out the danger and a growing animosity to barbed wire: "Farmers who live on lines of railways where barbed wire has been used extensively in fencing the roads are not pleased with the fence. They say that their cattle are injured by them and that their small stock gets through on the tracks."[53] At the end of a vivid 1880 story involving horses injured by barbed wire, the *San Saba News* heated up the anti–barbed wire rhetoric:

> The erection of a fence of barbed wire is a declaration of war; it is the expression of the farmer against the stockman; the pectinated wire speaks to the cow and says "I forbid you to exist." . . . [The cow] is frightened at a rattle snake; she shies against the fence; the serrated wire pierces her hide and she roars out a vain protest and rushes bleeding to a thicket where the fly comes, the blue bottle fly, with a buzz and a hum, armed with an ovipositor; the fly puts in those fatal eggs from which are hatched the worm called the screw worm "that dieth not, neither is he squenched" till the poor cow falls and dies, yielding her body to the buzzards and her bones to the mildew of the prairies.[54]

A newspaper account from the Austin, Minnesota, *Mower County Transcript* in 1876 was one of the earliest stories to point out what would become a primary objection to barbed fencing: the potential harm because of the fence's invisibility to animals, especially horses. "They do not see it as they do an ordinary board or rail fence, and consequently do not turn aside when they come to it. We have heard of a great many horses being injured by it, and some that have been completely ruined."[55] In Vermont, a series of newspaper stories appeared in rapid succession in 1879 that demonstrated a rising sentiment against the invisible fence. The *Burlington Weekly Free Press* argued that "the rapid introduction of barbed wire for fencing sufficiently indicates its efficiency. But a serious objection exists when it is used alone. Being invisible to animals at a distance, they have sometimes dashed hastily against it and had their flesh badly torn by the sharp barbs."[56] And the *Green-Mountain Freeman* punctuated its complaint with a pun: "We recently saw a very valuable colt badly cut across the neck, breast, shoulders and sides by becoming entangled in one of these fences. . . . For our own part, we would never fence our own animals in with such a *barb*-arous style of fence."[57] An 1879 issue of the *Lewiston Teller* reported complaints against "the barbed wire fence where no rail, pole or board is used. Posts are set about 20 feet apart and two strands of the barbed wire stretched horizontally. The stock see an opening between the posts, but do not see the wire, and often run upon the wire, and in some instances do much damage to themselves."[58] In the May 16, 1879, "Editors News" column, the *Emporia News* asked whether farmers could not "place with the wire at least one board as a safeguard to stock and stock men? Horses and colts are apt to run headlong, and unless there is a more apparent bar to their progress than an almost invisible wire-fence much damage must accrue."[59] The Jackson, Mississippi, *Weekly Clarion* continued this argument the following year: "Our attention has been called to the fact that the barbed fence wire . . . is one of the most effective agencies in existence for the destruction of life and limb of cattle and horses. . . . If farmers must use this wire, let them have at least a top board on the fence, so that an animal can see that there is a fence."[60] And in 1882, one of the Ohio *Jackson Standard*'s reporters made a strident call for action: "Stock would scarcely see these wires, and might run against them and be terribly lacerated. Here is work for Senator Brigham. Put railroad officials in the Penitentiary who put up these dangerous wires."[61] Reporting on wire fencing's invisibility forced proponents and opponents alike to address the negative impact on the very animals it was designed to control and to secure. And as the press continued to raise these questions about barbed wire, other problems came to light.

As early as 1879, the *Green-Mountain Freeman* reported on a lightning strike in Texas: "The big story Texas contributes is that lightning struck a fence of barbed wire, and after running along it for a mile, killed a young fellow named McPherson."[62] And from Deer Lodge, Montana, there was an 1881 report: "Last week one of our townsmen, who has an extensive ranch inclosed [*sic*] with barbed wire fence, found three cows lying dead by the fence, that had been killed by a stroke of lightning running along the wires. This is a new objection to inclosing large areas with wire fence that has not been anticipated."[63] And from Emporia, Kansas, that same year: "Complaints are made all over the West, particularly in Kansas, Nebraska, and Iowa, that wire fences attract lightning, which often results in the killing of stock. The stock imagine that the fence affords protection from the storm, and accordingly huddle up to it. One instance is given where sixty-five head of sheep were killed in Nebraska at a single stroke."[64] An 1881 story in Minnesota's *New Ulm Weekly Review* included an ominous warning for its readers: "Beware of wire fences. A few days ago one on the farm of Mr. Alexander, near Northfield, was struck by lightning and four valuable horses in the pasture near by were instantly killed."[65] Stories on the interplay between lightning and barbed wire multiplied. From Brenham, Texas, in 1882: "Six beeves were killed by lightning Tuesday night, on Mr. Wade Smith's place on Onion Creek, ten miles south of the city. It appears that the cattle were standing near the barbed wire fence, and the electric current struck the wire, splitting the posts and killing the beeves as above stated."[66] And from Owosso, Michigan, in 1883: "Seven steers standing beside a barbed wire fence near Decatur, Illinois, were killed by a strike of lightning, Monday afternoon."[67] Lightning strikes on barbed wire happened often enough that in 1882 the Washington, DC, *National Tribune* noted that many cattle and sheep in western states had been electrocuted while lying close to barbed wire fences in lightning storms and provided advice for their readers: "It is said that a good preventive can be had by twisting a good sized wire around the barb-wire at intervals not exceeding half a mile each, and running the lower end into the ground a foot or more, so as to reach the moist earth."[68]

A somewhat surprising addition to the litany of destruction came from tanners and leather workers. In 1883, the Napoleon, Ohio, *Democratic Northwest* wrote that "cattle on the grazing lands cannot see the wire, and . . . the hide is sometimes so badly scarred and cut that it is rendered almost worthless for leather where the grain is of consequence. . . . There is really consider[able] alarm among tanners on this account.[69] The *Prairie Farmer* ran a similar story three years later, noting that the prominent Chicago firm of A. F. Risser & Co. "regrets the deterioration of leather by injuries to hides caused by barbed wire fences."[70] While tanners were troubled by the defacement of cattle hides, the American Humane Society saw the issue from quite a different perspective. The *Las Vegas Daily Gazette* reported on a visit by an agent for the American Humane Society of Boston, C. L. Heywood: "From Mr. Heywood's valise was produced a section of tanned cowhide which showed a perfect net-work of cuts and slits made in the animal's back by the barbs of wire fence. It does not seem that such a trifle as a barbed point could make so many severe injuries and the GAZETTE is not surprised that the humane societies have had their attention called to the matter."[71] And from the Washington, DC, *Evening Star*, another story regarding the American Humane Association's interest in barbed fencing: "The American Humane Association . . . convenes in this city next Wednesday. The association considers national questions, such as cruelty to cattle in railroad transportation, branding live stock, the barbed wire fence."[72] The damage caused by barbed wire thus created an unlikely bond between those who advocated for animal welfare and those whose profits relied on the killing of animals. As the fourth estate continued to debate the merits of the new technology, it quickly became clear that animals were not its only victims. Humans themselves experienced the vicious nature of the "Perfect Fence."

Injury to Humans

One of the earliest references to potential barbed wire danger for humans appeared in the *Ottawa Free Trader* in the summer of 1876. The article describes a new city reservoir fenced with barbed wire as an effective way to "prevent accidents," because "very few will risk tearing their clothes and flesh to get a foot near the pool."[73] The *Weekly Kansas Chief* warned readers in 1878 that "it will be good policy for those who keep late hours, to be very cautious about a contact with Dr. Sawyer's barbed wire fence. A collision with said fence will create a feeling strange, painful, and alarming."[74] And in an article titled "Clear the Sidewalks" the Austin *Weekly Democratic Statesman* of 1882 warned its readers about possible injury: "In front of a hardware store on East Pecan street there is a lot of barbed wire stacked up, that has been there for months. If a person should go along there and step on an orange or banana peel, or fall down on this wire, severe injury would be inflicted."[75]

Barbed wire had a special affinity for clothing, as a series of accounts from 1882 and 1883 reveal.

> [Minnesota] The Lyon County *News* says a Fairview granger attempted to jump a barbed wire fence and came to grief in the operation. The slack of his new ducking overalls caught on the top wire, and before they gave way the granger realized exactly how it feels to be crucified.[76]
>
> [Louisiana] The new barbed wire fence in front of the Thompson property brings itself into notice by accidentally coming in contact with a fine piece of goods worn by one of our promising young lawyers.[77]
>
> [Illinois] Two young ladies named McCarthy, living several miles from Independence, while returning from the circus, were struck by

> the storm, their buggy overturned, and they blown against a barb wire fence. Their clothing was torn off and they were seriously lacerated.[78]

> [Kansas] Leavenworth ladies are indignant at hardware merchants for leaving coils of barbed wire on the sidewalks. One lady who spoiled her new lawn dress, is especially indignant.[79]

These mostly benign and sometimes humorous press observations had more serious counterparts.

The *Chariton Courier*'s "Haps and Mishaps" column for January 25, 1879, included a story from Iowa with a tragic outcome: "S. Johnson, an old resident, took down a barbed-wire fence to make a short cut home, when his team started suddenly, pulling him against a wire with such force that an artery in his leg was severed and he bled to death."[80] In 1883 the *Omaha Daily Bee* printed a column under the sensational heading "The Brutal Barb: A Woman Fearfully Cut Up by Being Thrown against a Barbed Fence." Mrs. Charley Howard, the account states, was left to drive a wagon while Mr. Howard drove some cattle home from a sale. The horses bolted, and

> Mrs. Howard [was] thrown against the wire fence, and dragged for some distance along the barbed wire. When picked up she was unconscious and remained for two hours. Her left arm was broken and almost cut in two by the wire, the collar bone was turned loose from the shoulder-blade on one side and her body and limbs were cut and bruised in a frightful manner in many places.[81]

In that same year, both the *Mower County Transcript* and the *Phillipsburg Herald* reported deaths: "Mrs. Martin Summers, at Vincennes, Ind., on the 19th was killed by being thrown on a barbed wire fence by a run-a-way team."[82] And from Kansas: "A lady of this place met with a terrible death a short time ago. The team which she was driving became frightened and run [*sic*] away, threw her on a barb wire fence, cutting her throat in such a manner that she died instantly."[83] The *National Live-Stock Journal* carried a gruesome story in 1888 about a man named Sassman who was chasing a stray steer: "It was just about dark, and as he went along after the animal he ran into a barbed wire fence, a barb entering his mouth and becoming firmly imbedded in the roof of it. Sassman was moving pretty rapidly, and his mouth was badly torn and jaw fractured."[84]

Stories of injured children were especially poignant: "A LITTLE boy, son of Mr. J. P. Strother, living near Independence, Mo., was killed last Tuesday by being thrown from a sulky-rake against a barbed wire fence."[85] "A daughter of John W. Wyman, East Liverpool badly injured her eyes, perhaps ruined them for all good, by running against a barbed wire fence, the other day, not seeing it. She is under the physician's care, and the result of the accident is much dreaded."[86] The *Wichita City Eagle* reported on a terrible case involving a two-year-old girl named Bessie: "She was out playing with her little brother very near the house in a lot enclosed with a barbed-wire fence, and in attempting to crawl between the wires her chin caught on one of the wires of the fence which resulted in strangulation by hanging in a very few moments."[87]

Simply stringing a barbed wire fence could be dangerous, as two accounts from Kansas illustrate: "Frank Hines has lost an eye. He was stretching barbed wire, when it broke and one of the barbs cut the ball of his eye open."[88] "Eli Sams caught hold of a barbed wire to see if it was stretched enough, when the wire broke and a number of barbs were instantly drawn through his hand, mangling it badly."[89] Descriptions of human encounters with barbed wire were common nationally well into the 1890s. In 1892, the *Chicago Daily Tribune* reported on yet another human–barbed wire encounter in a story titled "Climbed the Fence: Three Girls Escape from the Erring Woman's Home." In this case barbed wire had thwarted an escape when two of five escapees were spotted by a policeman passing the grounds. Officer O'Shea "noticed two dark objects hanging on the six-foot fence. . . . He crossed over to examine them and saw the girls, their clothing having been caught on the barbed wire nailed to the top of the fence. Thus entangled they were unable to either reach the ground or climb back over the fence."[90] Implicit in these kinds of stories was a plea for manufacturers of barbed wire to respond to the dangers their fencing posed to humans (unless, as in the Erring Women's Home, the barbed wire was meant to cause harm to escapees); and as discussed in the following chapter, there was some attempt to do so.

In addition to the accidental encounters with barbed wire, there were deliberate interactions that received press notice. An Idaho newspaper reported, for example, on the "Henry C. Champlin suicide at Jackson, Mich., by wrapping a barbed wire around his neck and suspending himself in a sitting position from the top of the fence."[91] In 1886 the *New Ulm Weekly Review* reported that among the more than six thousand suicides recorded nationally since 1882, an unusual method had appeared: "Still others clumsily sought death by crawling back and forth through barbed wire fences, entailing great suffering, until they died."[92] Whether intentional or accidental, the human encounters with barbed wire recounted in the press reinforced people's perceptions of the barbarous nature of this fencing. Nonetheless, because barbed wire fences were economical and highly effective, they continued to spread. Newspapers and magazines were left with an ongoing ambivalence concerning the barbed wire question.

Editorial Ambivalence, Vigorous Debate

As newspapers reported on the positive and negative effects of employing barbed wire, some agrarian magazines like the *Cultivator & Country Gentleman* were uncertain whether or not to promote barbed wire. In an 1877 article titled "Barbed Fencing," the *Cultivator & Country Gentleman* focused on the fence's invisibility: "We cannot quite approve the practice of making fences exclusively of this barbed wire fastened to posts. Being nearly invisible to animals, there is danger of their dashing against them, and becoming seriously injured by the barbs."[93] The *Cultivator*'s article ignited a firestorm among its readers. In a February issue, L. B. Colora from Maryland asked the *Cultivator & Country Gentleman* to make clear that "the dangers of the barbed wire must be apparent to every one. . . . I should think no man in his senses would be willing that his active horses should learn to respect such a fence by a personal encounter."[94] In a March issue, J. L. Carter contradicted Colora's claim: "The writer . . . says the 'barbs are positively dangerous.' I have never heard or read of an accident from them, although barbed wire fences have been in use for several years." In that same issue, another reader responding to Colora's letter wrote that he was "glad to know that the danger to stock from the use of barbed wire being at last pointed out. I have heard of some dangerous accidents to valuable stock in my neighborhood."[95] In April, a reader argued that barbed wire could be used safely: "Should I wish to turn into an enclosure of barbed wire an animal that is unacquainted with them, I lead him up to it and around it, letting him feel of it occasionally, and it will be perfectly safe to leave him in."[96] By June 1878, the magazine seemed to have had enough debate. When a reader from Canton, Mississippi, made inquiries into the "demerits as well as merits" of barbed fencing, the magazine responded tersely: "The subject has been already well discussed."[97]

Nonetheless, the ongoing debate about barbed fencing forced the *Cultivator & Country Gentleman* back into the breach. In 1880, the magazine published a series of letters from one of its regular correspondents, a William Horne, veterinary surgeon, who laid out the problems with barbed wire in no uncertain terms. In April he wrote that although barbed wire has its virtues, "it is unquestionably the most dangerous and cruel fence ever invented. . . . It is a murderous, treacherous snare, and inflicts much suffering upon our domestic animals. I am quite in a position to know whereof I speak. I have seen numbers of young colts literally torn to pieces."[98] In June he added that "I would like to ask any advocate of this murderous cruel fence why barbs were added to the wire, if not with a positive intent to be cruel, by the infliction of pain and injury to domestic animals. . . . What then is it used for, I would ask, if not for torture and ruin to some of our best and most costly animals?"[99] The *Cultivator & Country Gentleman*'s readership responded to Horne's pleas by submitting information on various devices designed to warn animals. W. I. Chamberlain of Summit County, Ohio, for instance, reported that a modified form of the fence was used in his region, four wires with the second from the top outfitted with a barbed wood strip: "This strip distinctly shows cattle and horses there is a fence, pricks them if they try to feed between the wires, but swings on the wire to which it is attached, so as never cruelly to lacerate an animal."[100] A more drastic solution was suggested by a Missouri reader who wrote that only one barbed wire fence had been put up in his vicinity, "and before its completion a valuable young horse was killed and other stock injured. The result was the cutting of the wire into short pieces by some unknown parties during the night. And the owner has at least for the present, abandoned the notion of a barbed wire fence."[101] Other readers responded to Horne's letters by enlisting logic in their support of barbed fencing. Suel Foster from Muscatine, Iowa, addressing his letter to the fence cutters in Missouri, argued that "there are as many persons injured by reaping, mowing, and threshing machines, as there are animals injured by barbed-wire fences," but farmers do not abandon those useful tools.[102] J. W. C., from Battle Creek, Michigan, claimed that owners of injured animals are at fault rather than the barbs, "which are assumed to be so barbarous, but are really preventive. . . . Barbed wire makes the best possible cattle fence . . . because it hurts just enough to prevent the destruction of the fence without materially injuring the cattle."[103] Horne's swift response to this last letter chastised J. W. C.: "I do emphatically and most positively condemn the inhuman and cruel intent of those barbs. . . . Until some method is devised to prevent animals from being torn to pieces . . . this fence remains a barbarous nuisance."[104]

Early that same year (1880), the *American Agriculturalist* called for a more thorough review of barbed wire:

> More than 100,000 miles of Barbed Wire Fence have been recently erected. Whether this style of fencing is best, or desirable; whether it is too barbarous or not, for general adoption over our vast, timberless prairies at the West, and what forms can be adopted in our more thickly settled regions, for highway fences, for village lot enclosures, etc., will be the subject of examination and discussion in this and especially in future papers. . . . Many of our readers have the new barbed fences. We earnestly invite accounts of their experience and observation, *pro* and *con*.[105]

The magazine was swamped with accounts and found it necessary to explain that "it is simply impossible to reply individually to many hundreds of letters. . . . The best we can do is to read over what is received and prepare for these columns such materials as seems to be most worthy of publication for general information." Proponents claimed that the benefits (low cost, availability, ease of installation, durability) outweighed any danger associated with barbed wire usage. And if animals did come into contact with the fence, they would quickly be repelled and less likely to encroach on the fence in the future. Detractors focused on injuries resulting from contact with the fence. In an attempt to undercut the outright advertising that began to appear in letters, the editors also asked readers to curtail their letter writing unless they had used barbed fencing for a year or so: "Several letters have come strongly recommending one kind or another, which have upon investigation been found to have come from *interested* agents, whose opinions are necessarily colored by such interest."[106]

In its March 1880 edition, the *American Agriculturalist* included a column titled "Injury from Barbed Wire" that laid out its position on barbed fencing. It noted its initial "strong prejudice against Barbed-wire fences." It endorsed the more visible Brinkerhoff steel strap fence with modified barbs. And it had recommendations for those who used the more common barbed wire: "While the strongly-barbed wires may probably be advantageously used for the large pastures of the West and South-West, we do not advise their use generally where farms are divided into smaller fields and where valuable animals are kept."[107] Responses came quickly. H. L. Raven of Travis County, Texas, claimed, "As to the injury to live stock, which seems to be the great objection to barbed wire here, my own experience is that the danger is not of great extent. I have had no serious accidents. Only one horse has scratched itself."[108] R. Noyes, of Coles County, Illinois, commented, "As to its being 'barbarous,' I have never known an animal really hurt with it, and if they are scratched they will not try it again." Warren Stone of Lyon County, Kansas, wrote that "here on these treeless prairies barbed wire is almost a necessity, and without doubt is a great blessing. . . . Almost every kind of barbed wire is in use here, and none are too severe; the town cattle will stop for nothing else." James R. Gillis of Louisa County, Iowa, reported that "I am using barbed wire fencing, and find it efficient . . . in restraining cattle and horses. . . . I have used barbed wire fence for some months as a pasture fence for ten horses, and have only had one scratched, and that not seriously."[109] R. A. Steele, of Douglas County, Kansas, wrote about his conversion to the new fencing: "I have been using barbed-wire for the past year; was previously prejudiced against it; have now over one mile. . . . It answers the purpose; that is, it turns stock. . . . I prefer the four-barbed rather short; the long ones cut severely."[110] And a paraphrased letter from R. C. McWilliams of Northumberland County, Pennsylvania, to the *American Agriculturalist* expressed a more nuanced opinion: "He [McWilliams] has not discarded its use, and does not absolutely condemn it; hopes the embankment described last month . . . or some other device, will render it less dangerous. He had a valuable cow that had one leg cut down to the bone, 'nearly half off,' and the flesh badly torn by the fence barbs."[111] Even as he noted the dangers of barbed fencing, L. B. Goodwin from Rock Island County, Illinois, claimed that horses would learn to avoid the fence: "I know of several instances where horses have been seriously hurt by running against the barbs. . . . I consider barbed wire as dangerous to stock unaccustomed to it, but as cattle and horses need but one introduction to the stranger, and shun it ever after, we can not afford to dispense with it." Louis Voruz from Coffey County, Kansas, echoed Goodwin's opinion: "I have had a year's experience with one kind of barbed fence. . . . As a rule, an animal which has once been through that fence does not care to repeat the experiment." Voruz added that he was in "favor of sharp, straight barbs . . . *but not long enough to penetrate the vital part of any living thing*."[112] A reader from Wisconsin urged the employment of an even more vicious wire: "As to the present forms of barb wire being dangerous to stock, I have only to state that whoever says so, does not know what he is talking about. . . . There is not a single form of Barb Wire at present in use that is half *barbarous* enough. . . . In fact we need a more severe barb wire introduced."[113]

The debate over the qualities of barbed fencing forced the *American Agriculturalist* (as it did the *Cultivator & Country Gentleman*) to revisit the topic and its position three years later and to take a definitive side in the debate:

> Steel wire in some of its many forms is becoming, in certain quarters, the fencing material for farms and railroads, and even for garden and lawn. Some of the points claimed in favor of this new fence are: cheapness, durability, ease of shipment of the material, and indestructibility by ordinary fires, like those arising from sparks of a passing locomotive. . . . A colt, or cow, that has once been pricked by the barbs, will keep a safe distance from the fence. Most of the damage has been done by the animals running upon the wires without seeing them. A strip of board may be nailed from post to post, near the top, to make the fence more conspicuous.[114]

This was a significant change from its ambivalent position in 1880, one perhaps swayed by its readers' comments.

Defense of the new fencing also appeared in newspaper editorials across the nation. The *Lancaster Daily Intelligencer*, a strong proponent of barbed wire in its editorials and its popular "Farmer and Gardener" column, noted,

> These fences are, so far as we see and learn, gaining in popularity. The objection to them at first raised, of their being liable to injure cattle, is practically without foundation. The spikes or barbs are too short to do any real harm. At the worst they could do nothing more than scratch an animal. . . . This fence possesses other advantages besides merely turning horses and cattle. It will do the same for dogs and biped trespassers, which will prove very serviceable.[115]

The following year, the *Daily Intelligencer* again wrote at length on the value of barbed fencing:

> The objections urged against the general use of the barbed wire fence, so far as our observations extend, have very little weight and will exercise very little influence against it. . . . We doubt if an instance can be produced where a farmer has abandoned it after a fair trial,

> in consequence of any damage it has done or on the score of expense . . . and a few years more, as wood becomes scarcer and dearer, the barbed wire fence will not merely be a matter of necessity, but of choice, alike for its efficiency and advantages.[116]

Similar sentiments were expressed in many newspapers, for example, in the Abbeville, Louisiana, *Meridional* (1881); the Point Pleasant, West Virginia, *Weekly Register* (1881); and the Montana Territory's *Rocky Mountain Husbandman* (1883).[117]

Other newspapers expressed their unambiguous opposition to barbed wire. The St. Paul *Daily Globe* printed a series of editorial comments over several years demonstrating its unshakable opposition, of which these two are representative: "Mr. Jebb, of Union Grove, had a young horse cut himself up very badly on a barbed wire fence. Judging from the number of accidents of this kind mentioned in nearly every state paper, these barbed wire fences are a source of great trouble to horse owners."[118] "We should rather say that persons should never put up the barbarous barb wire fence. A more vicious, cruel, dangerous and abominable invention for a fence was never made. Its use ought to be wholly discontinued."[119] In an article titled "Barberous Wire Fence," the Winnsboro, South Carolina, *News and Herald* concluded that "the almost universal opinion is that it is objectionable on account of its severity . . . and very liable to injure stock."[120] A disgusted Prescott *Weekly Arizona Miner* noted in May 1881 that "the man who invented or uses barbed wire for building fences, ought to be compelled to wear a suit of underclothing made of barbed wire cloth. The one is the refinement of cruelty to animals, and the other would only punish the human animal for his cruelty."[121] The *Worthington Advance* exclaimed, "There ought to be a heavy penalty against using barbed wire fences."[122] The Butte *Weekly Miner* noted, "There is much complaint about barbed wire fencing. . . . Those who use it should do so in a manner that will not inflict damage, or they will render themselves liable to the penalties of the law."[123] The St. Johnsbury *Caledonian* simply noted, "Wire fences are bad things and should be ruled out."[124] The *Saline County Journal* resorted to hyperbole: "Wire fences are death dealing, infernal machines and should be wiped from the face of the earth."[125]

The public clamor and debate about dangers posed by barbed wire fences led to reports of legislative hearings on the question. This piece from the *Richmond Democrat* describes a bill that "provides that all barbed wire fences shall be constructed with a plank or rail addition, to be placed not less than two and a half feet from the ground. It is well known that until stock become familiar with these wire fence[s] they are constantly running into them and receiving severe injuries."[126] The *Mexico Weekly Ledger*'s "The Legislature" reported on the proposed requirement for a board addition for the purpose of visibility, described the arguments by opponents of barbed wire in general, and then cited the winning argument: "It was argued that its economy and adaption to pasture and sheep enclosures outweighed all that could be said against such a fence. The recommendations were tabled and the matter thus disposed of."[127] Newspapers like the *Pascagoula Democrat-Star* advocated state action: "The cruel barbed wire fence is used largely in the Yazoo valley. Some day our Legislature will be called upon to forbid its use, under the 'cruelty to animals' statute."[128] And others, like the *Dallas Daily Herald*, made multiple calls for state action: "In all our prairie regions timber is scarce and there is no recourse but to hedges or barbed wire. Hedging is a long and tedious process, and the barbed wire ought to be prohibited by law. It is both cruel and dangerous."[129] And then again: "It is a convenient fence, it is true, but a most cruel one, and we doubt if in cattle grazing districts its use ought to be permitted at all. If we were a legislator we should most assuredly urge the passage of a law prohibiting its use in the cow counties of the state."[130]

The *Abbeville Press and Banner* printed an appeal to cattle owners: "Don't build barbed wire fences. The cattle cannot see the wires, and running against them, seriously injure themselves. There should be an act of the Legislature prohibiting the erection of any such dangerous devices."[131] The *Milan Exchange* questioned the state's approval of barbed fencing and the capabilities of farmers who employed the wire: "We doubt the propriety of the legal fence in Tennessee, established by our late Legislature, whereby the *barbarous* barbed wire is made to fill the requirement of the law. Such fences should be positively interdicted by law for all public highways. . . . Farmers who cannot take care of their own stock, had better not have any."[132] And in 1884, before the worst of the blizzards that caused the loss of 65–70 percent of herds trapped against barbed wire drift fences in the Texas Panhandle, a statement by the Colorado Cattle Growers Association laid out its opposition to barbed wire as a "veritable instrument of death" in both winter and summer conditions: "And even in the summer time they [fences] are not entirely, free from danger, for we notice that very recently a herd was stampeded by a hail-storm and came in contact with a wire fence, along the line of which the older and heavier cattle trampled to death 300 of the younger and weaker ones."[133]

Cattlemen and farmers from across the nation responded to the growing animal death tolls by tearing down many of the barbed wire fences they had set up in previous years, especially drift fences, replacing them with woven wire or older types of fencing.[134] Barbed wire was effective in containing and slowing the drift or stampede of cattle; but perhaps it was too effective. And while these aspects of barbed wire were being debated, another set of objections arose, these to the specter of monopoly.

As historian Charles Geisst finds in his study of late nineteenth-century industrialization, the public "tolerated [monopolies] as long as they produced fair prices." But when industrial giants began to reap exorbitant profits, public outcry arose. Geisst argues that industrial empires that make "tangible products with clear lines of supply" were "easy targets" for antimonopoly factions who could put a face on a corporation that could be attacked.[135] Barbed wire was one such "tangible product." Jacob Haish, whose patents were being challenged in court, skillfully orchestrated challenges to the monopolistic threat posed by the Washburn & Moen Company. Although an article in an 1878 edition of the *Weekly Chillicothe Crisis* did not identify Washburn & Moen by name, it did set up a "David versus Goliath" story that pitted Haish against the monopolistic giant:

> The introduction of barb wire fence . . . threatened the farmer with extortion through an assumed monopoly. Patents were secured by certain parties which were supposed to control the entire manufacture of the barb fence. At about the same time Mr. Jacob Haish . . . determined to proceed at once on the right common to all men to manufacture, and farmers may thank Mr. Haish for his pluck, his perseverance and his sagacity, and for breaking the extortionate prices. . . . No man, or combination of men, enjoy a moral right to impose by extortion. . . . Skill and genius deserve reward, but when the price demanded assumes the demand of a Shylock it is legally the privilege and duty of the public to resist.[136]

The article closed by encouraging its readers to take a "determined stand on the side of Mr. Haish" and implored farmers and dealers to use Haish's fence. As Washburn & Moen's lawsuits worked their way through the legal system, more articles appeared that continued the tone set in the *Weekly Chillicothe Crisis* story. In Minnesota, the *Mower County Transcript* warned near the end of 1878 that "now there is a report that the inventor of the Glidden barbed wire has won a suit establishing his right and title to the barb business and all infringers and all users of other makes of barbed wire are to be hauled over the coals and made to suffer."[137] In the spring of 1879, the *Weekly Chillicothe Crisis* reaffirmed its position of support for Haish against the monopolistic Washburn & Moen: "We advise all who are now using or who contemplate using Haish's Patent Barbed Wire to do so and to rest easy and keep the plow going, for when Jacob Haish says he will protect you it means business!"[138] In the wake of the December 1880 *Blodgett* decision in favor of Washburn & Moen and against Haish, the *Canton Advocate* of the Dakota Territory ranted at length against the moneyed interests of Washburn & Moen and the federal patent system that permitted the barbed wire giants to acquire a legal monopoly: "These chaps [Washburn & Moen] conceived the brilliant idea of buying up all patents granted to barbed wire inventors, and thus monopolize the enormous business that was growing up. . . . They have thus managed . . . to grasp and control absolutely one of the greatest industries of the country."[139]

If this was not enough to inflame readers' sense of fair play, the *Advocate* made dire predictions of things to come as result of the Yankee monopoly:

> They can dictate terms to all manufacturers, hunt them down, fix their own tariff, and put as many millions in their own pockets as they choose to. When they get through this extortion . . . they can organize a grand army of spies to hunt up and tax every farmer who has a barbed wire fence. . . . These shrewd Yankees do not claim to have invented the barb or anything else that is useful or valuable. They have only invented this grand and sublime grab, that will put millions in their pockets that they have never earned.[140]

In an 1881 *Omaha Daily Bee* column titled "The Farmers' Alliance," Allen Root, the secretary for the new alliance, recounted Washburn & Moen's legal victories and reiterated the *Advocate*'s predictions. Three Massachusetts men, Root asserted, have been securing "all patents relating to barbed wire for fencing. The supreme court has just decided all contesting suits in their favor. . . . There is nothing now in the way to prevent this monopoly from laying just as heavy a tax upon the farmers as they see fit."[141] This will happen, Root warned, unless farmers "combine to protect themselves against being imposed upon by combinations." He claimed that the formation of a labor combination (like the Farmers' Alliance) was the only way to break Washburn & Moen's hold on the barbed wire industry: "Combinations can only be met by superior combinations of stronger force." Individual farmers would not be able to fight against the industrial giant, but "let the farmers associate themselves together throughout the state and determine not to purchase a pound of barbed wire for the next four years [when the patents expire]. Let the other western states do likewise."[142]

In April 1881, the St. Paul *Daily Globe* reported on a Des Moines farmers' meeting "in opposition to the barbed wire monopoly . . . with quite an attendance."[143] Farther west, in a column titled "To the Farmers of Montana," the Fort Benton *River Press* agreed that patent laws stimulate invention but asserted that the exorbitant profits being reaped by the monopolies amounted to "brazen-faced robbery of the consumer. . . . The farmers of Iowa, Nebraska, Missouri, Illinois and Kansas are organizing. . . . Stock-men and farmers of Montana should organize at once . . . and relieve the farmers of their greatest and most direct source of tribute robbery."[144] Similar editorial positions appeared throughout the West. In August 1881, the Dakota Territory's *Jamestown Alert* included a story about the Iowa Farmers Protective Association, which proclaimed that "the Washburn house will find that it has got hold of a set of people who don't scare worth a cent; whose motto is, 'Our liberty we prize; our rights we dare maintain.'"[145] In October the *Dallas Daily Herald* claimed, "If our farmers now stand firm, and fight it through to the end, there can be little doubt that the monopoly can be broken up."[146] And in November, the *Sacramento Daily Record-Union* stated, "It is to be hoped that the efforts to maintain so formidable a monopoly may be frustrated. No one firm or set of men should hold a monopoly over the future fence material of the world."[147]

Despite the journalistic attacks against monopolists and the rise of the Iowa Farmers Protective Association and the national Farmers' Alliance, Ellwood and Washburn & Moen were able to tie up their competitors in years of litigation that would be ended in 1892 only by a US Supreme Court decision.[148] Even the almost unanimous passage of the 1890 Sherman Antitrust Act (threatening punishment for anyone "who shall monopolize, or attempt to monopolize, or combine or conspire with any other person or persons, to monopolize any part of the trade or commerce") failed to challenge the spread of monopolies and their ability to limit their competition and control prices.[149] That meant that the corporate "Goliath" Washburn & Moen was still in firm control of the barbed wire industry at the end of the nineteenth century, having come to a settlement even with Jacob Haish.[150]

While the barbed wire business flourished—there was enormous demand for the fence—the newspapers that had employed such colorful rhetoric in defense of and in opposition to barbed wire also made sure that barbed wire worked its way into everyday discourse. The perfect (death-dealing) fence was a natural source of humor.

Barbed Wire Humor

In 1877, the *Weekly Kansas Chief* reported that "an amusing incident took place in town, the other day. Old Adam Muddy (colored) got fast on a fence, at Kuchs', that was made of Kelley's barbed wire, and he could not get loose till Bob Ray, another colored man, helped him loose."[151] The story's racial stereotyping of African Americans as naïve and bumbling in their encounter with barbed wire was also a staple of advertisements, as discussed in chapter 3. An account in the *Ottawa Free Trader*'s column "Cream of Our Exchange" features fighting women as the source of humor and barbed wire as the punch line:

> Two belles of Toddtown, Iowa, went to a ball with a joint beau; but he was not impartial in his attentions, and the fair ones went out to fight. "During the hair pulling which ensued," says a witness, "the contestants got up against a barbed-wire fence which appeared to be hungry for calico." Their clothing would now make a good variegated pattern for a rag carpet.[152]

The story of wrestling women whose amorous differences were resolved by the intervention of barbed wire was reprinted in newspapers from Tennessee to Michigan and Missouri.[153] Apparently, barbed wire and unrequited love went hand in hand, as in this story from the St. Paul *Daily Globe*:

> It was a very ingenious young man who was going to call on a young lady whose father kept a bulldog. He had endured several altercations with the dog, and had resolved that if the canine tasted of the back of his trousers that night, it would wish it hadn't. So he put a coil of barbed wire inside the garment to tear the animal's jaws. He went up to the house chuckling over his ingenuity, but the dog didn't trouble him further than to snuff at his heels. He was shown into the presence of the ladies in the parlor and asked to take a seat. That was a contingency

> he hadn't prepared for. He thoughtlessly sat down, but instantly arose in a manner so excited that the ladies knew something was wrong, and asked what was the matter. He thought best not to tell, but the haste with which he left the house and fled into the darkness convinced them that something was wrong, and when in the shelter of a friendly clump of bushes he removed the wire, he concluded that it would have been far better to have been bitten by the dog.[154]

And for wives who dragged their less-than-willing husbands to Sunday services, there was this bit of wit from the *Rock Island Argus*, a story reprinted in newspapers time and time again: "A lady says her husband will sit on a barbed wire fence all afternoon to see a base-ball match, and never move a muscle, but when he goes to church he can't sit in a cushioned pew for fifteen minutes without wiggling all over the seat, and changing his position forty times, and she asks how she can fix him so he will sit still in church."[155]

Barbed wire was often described as anathema to the "lazy laborer." "It is a noticeable fact," the *Chicago Daily Tribune* claimed, "that the people in the prairie country of Iowa do more work than any other people in the West. This is because there is nothing about the top rail of a barbed-wire fence that invites men to sit on it and talk politics while the grasshoppers get in the crops."[156] A Vermont newspaper conceded that "since our recent criticism of barbed wire fences, we have received an explanation which disarms complaint. One old farmer, who has barbed his whole estate, says that he gets one-fourth more work out of his hired men than he used to when he furnished a top rail to sit on."[157] And in Washington, DC, the *Evening Star* bantered that "There are two things in the world which a tramp will not sit down on. One is a barbed wire fence and the other is a good healthy hornet's nest."[158]

Newspapers also drew barbed wire into the political arena. In 1879 the *Dodge City Times* quipped that "if the barbed wire fence comes into general use, where is the undecided politician to seat himself?"[159] Politicians were singled by out by name in conjunction with barbed wire, as was the US senator from Illinois David Davis, whose name appeared in the *Butler Weekly Times* and the *Omaha Daily Bee*, respectively: "David Davis says a man will never sit long on a barbed wire fence unless there are bad dogs on both sides of it."[160] "David Davis's exceedingly bad humor since his return home induces the belief that during his visit to his mother-in-law in North Carolina he must have roosted astraddle of a barb-wire fence."[161] The *Shenandoah Herald* printed a tongue-in-cheek response to a lengthy 1883 congressional debate: "No wonder the Senate reduced the tariff on barbed wire. Every person who ever climbed a fence made of that material well knows that barbed wire is fully able to protect itself."[162] And more generally, the *Middlebury Register* reported that "Philip Stone has removed his dooryard fence and replaced it with a handsomely arranged barbed wire fence, on which he will be happy to see people sit and discuss politics and the weather."[163]

In a nod to the ongoing barbed wire patent disputes between Washburn & Moen and nearly every other barbed wire company, the *Ottawa Free Trader* weighed in with a sly bit of wit: "There's been considerable talk lately about 'royalty on barbed wire.' All we can say is that royalty must feel mighty uncomfortable."[164] Sometimes the victims of encounters with barbed fencing were the butt of the jokes: "Those parties who entered the residence of F. Brown, at Neosho Rapids, a few evenings ago, should be more careful next time. A roll of barbed wire is a bad thing to fall upon, for it leaves a mark."[165] The *Edgefield Advertiser* included this somewhat painful ditty in its August 31, 1882, edition:

> There was in the State of Ohiah
> A maiden named Helen Maria,
> Who ever would sail
> Down a banister rail
> When she thought there was nobody nigh her.
> Now her brother, whose name was Josiar,
> Fixed the rail with a piece of barbed wire;
> But it wouldn't be best
> To tell you the rest,
> For we are blushing already like Fiar.[166]

Leading questions from the *Sedalia Weekly Bazoo*'s column titled "What the Paper Published for the People Now on Earth Would Like to Know" wondered "if 'Pete' Merrifield has succeeded in getting his unmentionable half-soled, and in case he has, if he will steer clear of female seminaries hereafter where there is a barbed wire fence enclosing the premises?"[167]

Humans were not the only targets of barbed wire humor; animals also received their fair share in the press. In 1882, the *Helena Weekly Herald* included a quip reprinted across the nation in various forms: "Texas steers are more than delighted with the barbed wire fencing. They now have something to scratch on and wear the wrinkles off their horns."[168] Nor were Texas cattle unique in employing barbed fencing for grooming, as noted by the Breckenridge, Kentucky, *News*: "Breckenridge county cows are utilizing the barbed wire fences by using them for currycombs."[169] And, no doubt in response to the falling prices of barbed wire as a result of lawsuits and the Iowa Farmers Protective Association's move to produce its own antimonopoly barbed wire, the *Butler Citizen* reported: "Barbed wire fencing has fallen ten per cent in price within the last three months. Cows have got so they use it for a hair brush."[170] Cattle were not alone in figuring out advantageous uses of barbs, as noted in the *Emporia Weekly News*: "The American goat is gradually being narrowed down to a frugal diet of old crinoline and barbed wire fences."[171]

These vignettes were intended to elicit humor from a public that had become familiar with the qualities of barbed fencing. The stories work because of the constant presence of the prickly fence in newspapers over the last quarter of the nineteenth century. Through their descriptions and their various positions on the use of barbed wire, newspapers and magazines influenced the public's perception of the new technology. The constructed meanings eventually became rooted in people's daily lives, in their language, and in the literature they read. That barbed wire was so intensely and so publicly debated meant that the controversy could not be ignored by legislative bodies, inventors, or manufacturers. Responses by those groups helped solidify the meanings ascribed to the wire, as we demonstrate in the next chapter.

CHAPTER TWO

"Secure and Safe Alike"

Legislative Challenges and Inventive Responses

> If a cow thinks she knows better where she belongs than her keeper, the barbed fence has a tendency to correct her self-conceit. The world is still governed somewhat by the punishment of transgressors, though it shocks our humane feelings to have them hurt.
>
> — *Vermont Phoenix and Record and Farmer*, 1880

> The barbs gave the fence its power. They were about an inch long. The Washburn & Moen Company of Worcester was making the barbs shorter; he did not like that so well. If he couldn't have the barb he didn't want the fence.
>
> — *Vermont Phoenix and Record and Farmer*, 1880

The growing public outrage evident in the newspapers and magazines surveyed in chapter 1 brought various responses from state legislatures and counterresponses from barbed wire manufacturers. Texas, Connecticut, Vermont, New Hampshire, and other states introduced bills to restrict or outlaw the use of barbed wire. The Texas legislature, for instance, introduced three bills, two of which went into effect, making the use of barbed wire illegal or requiring that all barbed fencing be modified by fastening a board near the top of a fence to make it visible.[1] As state legislatures began to debate the use of barbed wire, the nation's largest barbed wire manufacturer, Washburn & Moen Manufacturing Company, appeared at legislative hearings and submitted lengthy printed statements to state legislatures, barbed wire distributors, and public media.[2] Our examination of cases in three New England states (Connecticut, Vermont, and New Hampshire) reveals the arguments made against the fence and the counterarguments made by Washburn & Moen to protect the image of its product. It is a classic case of the influence one corporate entity could bring to bear on state legislatures.

Legislative Challenges to the New Fence

In February 1880, the Connecticut legislature's Joint Committee on Agriculture convened a session in Hartford to hear public testimony on a bill to prohibit the use of barbed wire along highways and railroads. The bill was proposed out of concern that the New York and New England Railroad Company would use barbed wire to fence a new right-of-way. In attendance were proponents of the bill led by Representative Sanford, several representatives from Washburn & Moen, and other opponents of the bill, including farmers from Fairfield County.[3]

The *Hartford Daily Courant* reported that eleven witnesses for the opposition testified that they knew of no significant harm resulting from the use of barbed wire. D. W. Brooks, a florist, responded to a recent report that one of his employees had his ear torn off in an encounter with a barbed wire fence, testifying that the accident was in fact "a mere trivial one, the man running into the fence at headlong speed in the night. His face and ear were only slightly scratched." Opponents of the bill also described the value of barbed fencing; Samuel Nather, for instance, proclaimed that barbed fences were a "great boon to farmers."[4]

The committee next called representatives of the Washburn & Moen Company, who read from a twenty-eight-page prepared manuscript that began with a paean to manufacturing and with flattery aimed at the state of Connecticut:

> It is by these [hearings] your committee will weigh their duty in the matter, that no oppressive legislation or measure against any industry be allowed to spring from clamor or prejudice, and find its place on the statute books of a State, among all her sisters the most largely associated with the mechanical industries, and most largely interested in the expansion of all the legitimate interests of the manufacturer.[5]

The Washburn & Moen statement included an impressive compilation of views on fencing by agricultural authorities over a sixty-year period—information collected from state agricultural reports, agricultural societies, and farm publications. The statement laid out a brief but thorough examination of New England fence law dating back to 1643, a history in which Connecticut legislation had favored fences decade after decade. Accompanying this legislative history was a history of fencing that emphasized the increasing demand for effective, cost-efficient, and readily available fencing materials.[6] Washburn & Moen then argued that the bill currently under consideration suggested no reasons for the proposed ban. In fact, barbed wire fencing was

> put upon its public trial without open charges. . . . What is the worst that can be urged against Barb Fence? . . . No adverse fact or hostile impression or prejudice, would be likely withheld. One thing is noticeable on the testimony of all dealers and agents. Barb Fence is most popular and sells most readily in neighborhoods where it is in most familiar use and best known. . . . Barb Fence has grown to represent a very extended industry. . . . It has to do with a very wide and long confessed demand of the farmers and the people at large.[7]

Washburn & Moen continued by addressing the question of the day: Was barbed wire a cruel fence? So cruel that legislation was needed to protect the public? They argued that the farmers who used barbed wire would be the first to oppose its use if it was vicious in nature. "It is his own horses and herds that would be pricked, and maimed, and impaled, were such the inevitable or the customary sequences of Barb Fence. He would be sure to let it alone in future farm experiments. His observing neighbors would not need a state statute to keep Barb Fence from their own farms."[8] The company's representatives had made "careful inquiries" of their local agents and dealers throughout Connecticut, as to "accidents and injuries resulting from the use of Barb Fencing." What they found was "no case of substantial injury." Instead they heard "unreserved endorsement of Barb Fence." Would such endorsement be possible, they asked, if the farmers and stockmen of Connecticut found barbed wire fencing "cruel and barbarous?"[9] The legislature, they concluded, should ignore unsubstantiated public outcry and allow manufacturers and users of barbed fencing to determine its use.

The committee then called on proponents of the bill, of which only two were noted by the *Hartford Daily Courant*. A. Ganung, a farmer from Danbury, said the impetus for his opposition to barbed wire was that he had been *told* that "in the west cattle and horses had been greatly injured" by barbed wire, although he himself had no "personal knowledge of any injury resulting from barbed fence." The second proponent, Representative Sanford, said he himself knew of "no positive injury from barbed fence" but that he had heard that "farmers having it in use found it injured their cattle."[10] With no direct testimony confirming the claims that barbed wire was a cruel fence, the committee and both legislative houses eventually rejected the bill. But Washburn & Moen's victory celebration was brief. A more comprehensive challenge loomed when the State of Vermont drafted not one but three bills in opposition to the use of barbed fences within its borders.

In fall of 1880, the Vermont state Senate introduced a bill (SB 67), "An Act for the Protection of Domestic Animals." In the same session, the state House introduced two of its own bills (HB 214 and 259): "An Act for the Protection of Domestic Animals" and "An Act Prohibiting the Use of Barbed Wire Fences in Certain Cases."[11] According to the *Vermont Watchman & State Journal*, the bills would "practically do away with the use of barbed fences in the state."[12] Senate Bill 67 and House Bill 214 required that all stretches of barbed wire erected along a highway or between properties have a four-inch board fastened to the top wire, between all posts, to prevent domestic animals from injury should they run into the fence. Both bills included penalties for violations.[13]

House Bill 214 was introduced on November 6 by a Mr. Giddings of Bakersfield, read the first and second times, sent to the House General Committee for review, and then recalled and sent to the House Committee on Agriculture. By the December 11 session, the Committee on Town Lines had refused to hear a third reading of the bill. House Bill 259 met a similar fate, as the Committee on Town Lines dismissed the bill at its December 13 session.[14] The Senate's version of House Bill 214 (Senate Bill 67) was introduced, read the first and second times, and then referred to the Senate Committee on Agriculture in its October 28 session. By December 20, this bill, like the two House bills, was a dead letter.[15] What accounted for the rapid dismissal of these three bills restricting or prohibiting barbed fencing in the Vermont legislature?

Between the introduction of these Senate and House bills in late October and their subsequent demise by late December, the chairmen of the Senate Committee on Agriculture and the House Committee on Town Lines organized a public hearing in Montpelier concerning the three pending bills. The *Vermont Watchman & State Journal* noted that a large number of legislators and interested parties were present at the November 24 proceedings. Mr. Palmer, who had introduced House Bill 259, started the hearing by stating that his constituents did not object to wire fencing, just to barbed wire fencing. He spoke specifically of the Burlington & Lamoille railroad's use of barbed wire fencing in lieu of board fencing along its line through Jericho, Vermont, a measure to prevent snow drifting. Palmer stated that neighboring farmers "complained greatly of injuries to their horses and cattle in the pastures" bordered by barbed wire. He stated that in his town most people viewed barbed wire fencing as dangerous for stock and that it should be prohibited. Palmer's testimony was echoed by Senator Dana from Addison, who remarked that he had constituents whose valuable horses were seriously injured by the employment of barbed wire along highways.[16]

After these legislators' testimony, the rest of the hearing consisted of statements by farmers and others who claimed that barbed fencing was an essential tool for their work. Witness after witness, twenty-two in all, testified that not only was the fence effective, but they had no knowledge of an "authenticated case of injury" to animals or humans. A man named Goddard, speaking on behalf of Ludlow farmers in opposition to the bills, argued that barbed fencing was "a good deal softer thing to be thrown against than a stone wall." He commented that if perhaps "occasionally a horse or a cow . . . got pricked by it, there were hundreds that were prevented by it from straying away and being killed." So strongly held were his convictions that he was prepared to prove, he said, "that if the whole town was fenced with this fence, there would be far less accidents than with the old board and stump fences. Its use would save the State thousands of dollars in breaking roads, building fences, saving crops, and in the extirpation of vermin. It was one of the greatest inventions of the age."[17] Crosby Miller, a judge from Pomfret, began his testimony by claiming he had personally examined the "question of barbed fence." In doing so, he had never heard of one "authenticated case of injury." He continued with a story about his son-in-law who put up a barbed fence along his property line. At some point a neighbor's horse ran away and into the fence with such force that it took down one of the fence posts on impact. The horse was "scratched but not badly," and it recovered to "be just as good as he ever was." This episode "satisfied" Miller that "all this talk about wire fencing being cruel was moonshine." For this witness and many others giving testimony, the use of barbed fencing was "not a debatable question. . . . It was much safer and more efficacious than any fence known."[18]

In addition to direct testimony, ten written depositions were added to the hearing's minutes, all of which registered opposition to the pending legislation. At the end of the reading of these depositions, the *Watchman* reported Goddard's summation of the opposition's position: "The objection was that it was cruel. This was a prejudice and not a fact. The use of this fence would save many animals from being killed on railroads, and would do nothing more than scratch any stock. All these objections were based on hearsay."[19] When the proponents of the bills had nothing further to add, the hearing was adjourned. By December 13, the Vermont House Committee on Town Lines dismissed House Bill 259 and submitted its report to the Vermont legislature. The legislature overwhelmingly rejected the proposed measures, citing that barbed fencing was "not harmful" and "less productive of accident to animals" than other fencing materials.[20]

Seizing the moment, Washburn & Moen highlighted the outcomes of the Vermont and Connecticut legislative debates in their trade publications. They claimed that the farmers called to testify had provided no "adverse testimony, though every opportunity was given. As a result, Barb Fence stands stronger than ever before."[21] Washburn & Moen included a reprint of the entire November 24, 1880, *Watchman* article in an eight-page pamphlet added to the company's 1881 *Manual of the Fence*. In addition to the article reprint, the company included its take on what happened to the Vermont House and Senate bills. The legislature, it reported, "rejected by an overwhelming majority all the measures proposed as restrictive of Barb Fencing." The pamphlet continued with unabashed emphasis, asserting

> That it is NOT HARMFUL under any of the farm or roadway exposures.
>
> That it is a NEEDFUL material, since, with all the woodlands of the State of Vermont, there are thousands of acres that cannot be fenced without it.
>
> That it is used and may be used by all stock growers, with the certainty that it is on an average, LESS PRODUCTIVE OF ACCIDENT to animals than any style of old fashioned fences of stone, wooden rails, stumps or pole.
>
> That it cannot be burned down, torn down, blown down or stolen.
>
> IN OTHER WORDS THE FARMERS OF VERMONT LIKE BARB FENCE.[22]

This litany of self-praise was followed by a one-page "Recapitulation" that listed twelve statements about fencing in Vermont. The twelfth claim is of particular interest:

> That from this testimony of Vermont Farmers, Steel Barb Fencing must stand freshly affirmed and established as

> The Cheapest and Best of fences.
> The strongest and most imperishable of fences.
> The Safest fence for the protection of crops.
> The least harmful fence to all kinds of stock.
> Secure and safe alike for horses and cattle.[23]

Washburn & Moen ended its pamphlet by asking the "prejudiced reader" to carefully examine all the testimony given before the legislature and to "tell us if we have in the least exaggerated its meaning and value."[24] The company thus skillfully turned a potentially devastating attack on barbed wire into useful advertising for its product, a product that was effective because it caused pain and for that reason needed a positive spin.

Following the hearings in Connecticut and Vermont, Washburn & Moen faced perhaps its greatest challenge, this time from New Hampshire's State Society for Prevention of Cruelty to Animals. The society twice sought a restrictive measure from the New Hampshire legislature, forcing Washburn & Moen representatives to make multiple appearances before the state Committee on Agriculture before the issue was resolved in the company's favor.[25]

In their July 1881 statement to the New Hampshire legislature, Washburn & Moen addressed the question raised by the society: "Is it a Cruel Fence?": "Not unless a multitude of American Farmers who are using Barb Fence are blind to their interests and unmerciful to their beasts. Have not accidents happened from its use? Do they not happen from all fences?"[26] The company told the New Hampshire legislature that if it would "weigh rare and isolated cases of accidents against average and widely common benefits" it would come up with no strong rationale for the prohibition of barbed wire fencing. The statement continued with a brief account of how state legislatures in both Connecticut and Vermont had earlier attempted to outlaw the use of barbed wire but ultimately failed. These failures, the company reminded New Hampshire's statesmen, occurred as a result of "exhaustive" hearings that included testimony from the "best farmers in the State," whose own words established "more strongly than ever the merits of Barb Fence, and the unwisdom of adapting legislation to sparse accidents, which in nowise affect its universal utility."[27] After its initial investigation, New Hampshire's Committee on Agriculture unanimously refused to recommend the bill. Still, the legislature (perhaps under pressure from the society) sent the recommendation back to committee for another review. After a "full hearing" the committee again decided against restriction.[28]

Ironically, in its New Hampshire statement Washburn & Moen had introduced a further argument to reinforce the company's position that barbed fencing was not a "cruel fence." The company claimed that despite the legislative challenges, if there was merit to the claim of the Society for Prevention of Cruelty to Animals against barbed wire, surely it would have spoken out on behalf of "dumb brutes" and would have publicly censured the use of barbed fencing from its inception. That no such response had been forthcoming was proof, according to Washburn & Moen, that the society's own investigations into the alleged vicious nature of barbed wire turned up no substantial supporting evidence.[29]

Six months later the national Society for the Prevention of Cruelty to Animals publicly countered this statement by citing considerable evidence in their members' magazine, *Our Dumb Animals*, reprinting a particularly graphic article titled "The Barbed Wire Cruelty—a Nuisance That Needs Abating." The article included twenty specific examples from Pennsylvania counties and left no doubt about the editors' view that "the barbed wire fence is a cruel nuisance."[30] For instance, injuries "reducing the value and usefulness of the wounded animals, are not infrequent; but above and beyond these pecuniary losses is the excruciating agony which the poor, dumb brutes, rent and torn by the devilish barbs, are compelled to undergo." In one case, the article reports, "a dark bay mare cut herself across the throat from shoulder to shoulder. The cut was so deep that the workings of the windpipe were plainly visible, and the suffering of the poor brute was most intense." In another case, a horse "was chased by dogs, ran into the barbs and tore a frightful gash in his breast, which will disfigure him for life, and greatly diminish his selling value. Twenty-two stitches were required to sew up the wound." And in yet another case, a mule received injuries that convinced it thereafter to avoid the fence. "Experience is a dear teacher," the article moralizes, "but if some of our learners were endowed with the asinine qualities of the aforesaid mule, this extended list of accidents would not grow as the days pass by."[31]

Clearly there was merit to these Pennsylvania examples, but apparently not enough to alter the plans of corporate giant Washburn & Moen and its growing army of enthusiastic customers. As we have already seen in newspaper and magazine accounts, there *was* plenty of evidence from across the nation of the dangers posed by barbed fencing. Evidence of another kind can be found in a new industry that arose in response to injuries caused by barbed wire.

Healing Oils and Liniments

Injuries like those catalogued by the Society for the Prevention of Cruelty to Animals required healing. Not surprisingly, companies began advertising products meant to heal animals injured by the new fence. Prominently displayed in the Veterinary and Agricultural Department section of the 1889 Sears, Roebuck & Company's catalogue, for instance, one finds patent medicines such as the ACME Veterinary Wire Cut Remedy that claims to be the "remedy which should always be within reach."[32]

One of the most graphic depictions of the consequence of encounters with barbed wire is an advertisement for Silver Pine Healing Oil. Although the manufacturer claimed it was a "wonderful cure for barb-wire cuts," it was the image in the ad, rather than the words, that drew customers' attention. In the foreground stands a white horse with a large pool of blood forming at its feet from the parallel cuts inflicted by barbed wire. As shocking and painful as this must be, the horse itself does not appear to be in a panic. Instead, it waits patiently as a man rushes toward it with the "cure" in hand.[33]

The white, black, and red cover of *Haller's Catalogue, Health Guide, & Almanac* similarly featured the company's barbed wire liniment (fig. 9). In this image, a horse hopelessly entangled in four strands of barbed wire bleeds profusely as a small dog barks and runs between its legs.[34] The Scarless Remedy Company of Winterset, Iowa, emphasized the ugly scars resulting from barbed wire injury and produced a liniment whose label depicts two gentlemen farmers, an injured horse, and a seven-strand barbed wire fence in the background (fig. 10). The horse owner laments that his horse now "won't be worth a dollar," and his neighbor says that his own horse was cut even worse, but after using Scarless Liniment, "you can't find a scar on him."[35]

The labeling of barbed wire liniment for both "man and beast" was commonplace.[36] Although the barbed wire problem was a liniment seller's dream, the vicious efficiency of the fencing created a public relations nightmare for manufacturers.

Manufacturers' Arguments

Washburn & Moen soon expanded its appeals beyond the responses to legislatures. The company made its case in a series of trade journals. The American public, it claimed, employed over 250,000 miles of barbed wire in fields, on ranges, and along railroads. The sheer amount of barbed wire was proof

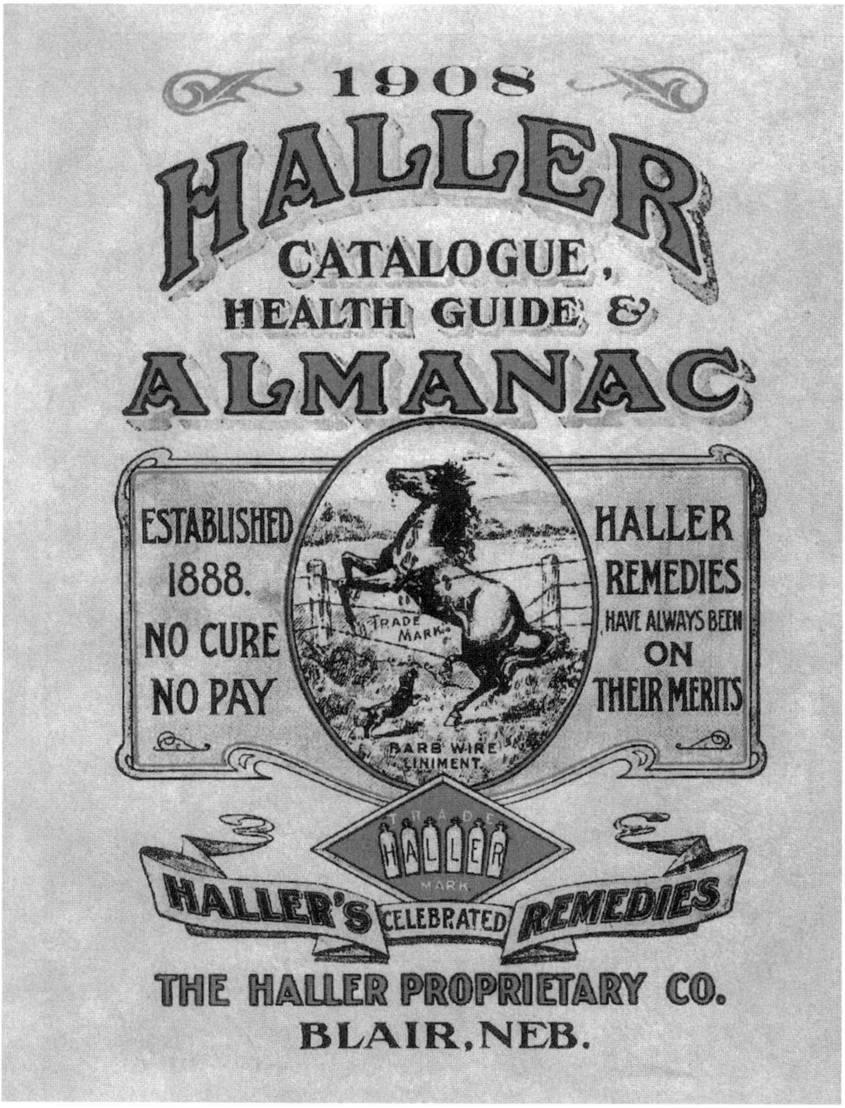

Figure 9. Cover of the 1908 *Haller Catalogue, Health Guide & Almanac*. Courtesy Kansas Barbed Wire Museum, La Crosse, Kansas.

of merit. In fact, "had evils attached to the use of Barb Fencing, the public would have been sure to hear of it, and this in loud and unmistakable terms. The objection to Barb Fencing comes in every instance from those who have had no experience in its use."[37] Washburn & Moen argued that its newfangled fencing was more humane and less cruel than fences that impaled animals on sharpened fence stakes and broke limbs of animals attempting to scale stone

Figure 10. Scarless Liniment label. Courtesy Kansas Barbed Wire Museum, La Crosse, Kansas.

walls. In contrast, "the accidents from Barb Wire have been mainly of a trivial character, in cuts and scratches, which, in such cases, have been warnings, salutary in their effect, and have educated the beasts in the new law of respecting fences."[38] The company addressed the rising clamor about the cruelty of its fence with a formal statement of "The Principles of Steel Barb Fencing." The "great central merit of Barb Fencing is in its capacity to inflict upon . . . an animal a sharp *instantaneous* prick." Without this ability to prick, wire fencing was ineffectual. A sharp barb was desirable so that it would "prick and repel at the moment of contact"; otherwise, "the *pain-giving* Barb Fencing" would be "a system for simply *scratching* the animal." The short, straight barbs avoided "tear[ing] the animal" or "hook[ing]" the animal that came into contact with it, thus lessening its potential for harm.[39] Jacob Haish echoed this sentiment in claims associated with his "S" Barb wire: It had "the cleanest, sharpest, shortest cut barb made. . . . The barbs are provided with a shoulder, preventing points from cutting and tearing the flesh. The only barb with a clear, sharp cut point, that will prick but not cut."[40]

Washburn & Moen emphasized that the barbed fence was a barrier "respected by the most unruly animals" and "employed by the best class of farmers," a fence "admirably adapted" in isolated fields to prevent theft and destruction, a fence that curtailed "cross-country practices" of "strollers and vagrants," a fence that commanded "instant respect from man and beast," but not by "wounding or positive injury."[41] To bolster its claim that barbed wire made an innocuous fence, Washburn & Moen pointed out that by 1880 sixty thousand miles of the "Perfect Fence" already surrounded fields, hemmed in railroad tracts, and bordered road rights-of-ways (scenarios illustrated throughout the manual).[42]

Another strategy was to argue that animals could and did learn the consequences of taking on barbed wire fencing from the moment of their introduction: "Animals acquainted with barb fences never get hurt by them, and it don't take them long to get acquainted."[43] It was simply a matter of "respecting the fence" on the part of the animal: "There are fewer breachy animals now than there used to be, and there will be less than at present when all animals are taught to respect the limits of the field by a good Glidden barb fence."[44] This idea that animals could be taught to "respect" the object of their imprisonment and then act accordingly, is recounted over and over again in the *Glidden Barb-Fence Journal*: "With barbed wire there is no such thing as a breachy animal. It positively cures the very worst subject and by enforcing its authority causing a wholesome respect for itself, none make even an attempt to pass the boundaries of the field, and hence by keeping out of the fire, none are burned."[45] Further, "No animal brought up behind a barbed wire fence ever becomes 'breachy.' With this fence they don't learn bad habits, and, as in other cases, the old adage, 'An ounce of Prevention is better than a pound of Cure,' holds good. But if bad habits are formed, barbed wire will cure them. It is both a *preventive* and *cure*."[46] In December 1881 the *Chicago Daily Tribune* also weighed in on the subject of animal education in an article titled "Barbed Wire for Fencing": "The first thing is to teach animals to entertain a becoming respect for a fence, and one that looks like a skeleton. While they are learning this lesson, quite likely some of them may get hurt; but once learned they always regard a line of posts as very significant."[47] Five years later the *Glidden Barb-Fence Journal* reasserted its claim that animals did eventually learn to avoid barbed fencing: "The theory was . . . that all animals on the farm were going to rush right upon it and commit suicide. The facts prove . . . that there are far less animals killed and injured on barbed wire than on the old style fences."[48]

The idea that animals could learn about the consequence of trying to breach a barbed wire fence, remember, and thus "respect" those encounters, and recognize the danger when it next encountered barbed wire, was a mainstay in manufacturers' literature and advertisements. An 1881 *Chicago Daily Tribune* editorial indicated that the arguments made by manufacturers were finding acceptance:

> Perhaps no device discovered or utilized since those of the reaping machine and the cotton gin has been of such value to the American people as the production of wire adapted to fences, and so provided with barbs, thorns and prickers, as to successfully exclude livestock from fields, or when in to keep them in. The cotton gin was applicable only to a single crop, and the reaper was also limited to grain, but barb wire fences are useful to every foot of the enclosed land of the United States, and to every man inhabiting a region where a horse, cow, sheep or hog, or other animal is produced or grown. It is useful where livestock roam, to keep them out of the cultivated fields, and as useful when they are herded to keep them in the allotted range.[49]

But enthusiastic reviews notwithstanding, simply stating that barbs were safer than other kinds of fencing and that animals learned to avoid them was not a sufficient argument for many. Perhaps there was another way to address the issue. Inventors responded quickly to the public outcry, modifying the injurious elements of barbed wire in what was eventually, despite prodigious efforts, a failed enterprise.

Inventors Respond to the Outrage

A comprehensive review of late nineteenth-century American barbed wire patents reveals that in the years from 1867, the year of the earliest patented barb in conjunction with a fence, to the last year of the century, 472 patents were issued for barbs, barbed fencing, and improvements to barbed fencing. Five patents were filed in the 1860s, 170 in the 1870s, 238 in the 1880s, and 59 in the last decade of the nineteenth century. The sharp increase in the number of patents granted in the 1870s reflects the rush of inventions inspired by the remarkable success of Haish's and Glidden's doubled strands of wire twisted to hold the respective barbs in place. The increase in new patents filed in the 1880s resulted from the quest to modify barbed fencing to address the various problems that arose with its application in response to public outcry and government intervention.

Henry and Frances McCallum determined that neither chronology nor design alone was the most effective way to analyze the changes in barbed wire across the last quarter of the nineteenth century. They argued that a more effective approach would be to focus on the inventor's purpose(s) in creating the wire. With this in mind, the McCallums discerned four general categories of barbed wire that fit within a loose chronology: (1) Early Varied (1868–1877), which included the earliest experimental and unrefined attempts at "patentable inventions," as well as the two patents that proved most successful over the subsequent decades: Joseph Glidden's "The Winner" and Jacob Haish's "S" Barb; (2) Vicious (1876–1880), where manufacturers embraced the ideas of longer, sharper, stronger, and more numerous barbs with the specific intent to "punish cattle that violated boundaries"; (3) Obvious (1879–1884), which included barbs but sought to quell rising anti-barb fencing sentiment in the face of livestock injuries by producing more visible fencing; and (4) Modified (1884–1892), which combined the best attributes of previous patents.[50] This system of classification makes good historical sense. An even closer reading of the patents' language reveals some interesting nuances that shed light on the shifting meanings of barbed wire.

Deterring Contact

A close analysis of the language of early patents discloses that control and learned avoidance were the primary objectives, or to put it more simply, the barbs added to smooth-wire fences would keep animals from pushing against and tearing down the fence. So, for example, William Hunt's 1867 patent, which attached a "series of small spur-wheels" to smooth wire fences, was designed to deter animals from "pushing against the fence or attempting to break over it." Michael Kelly's 1868 patent for a "thorny fence" claimed that it would "soon teach cattle to respect [the fence] and not attempt to force it." Henry Rose's 1873 patent stated it would "prevent stock from pushing against [the fence]." Jacob Haish's 1874 barb would "deter cattle and horses from throwing their weight against the rail," and Joseph Glidden's 1874 patent claimed that it had "means for preventing cattle from breaking through wire fences." Isaac Ellwood's 1874 patent said that "animals coming up to the fence will run against some of the points . . . and naturally go away from the fence." Charles Kennedy's 1874 patent attached "stars" or "protectors" to smooth-wire fencing and promised that cattle and horses would "soon learn to keep away from a wire fence." The Merrill brothers' 1874 "brads" would "prick the animal, and thus cause it to keep away from the fence." Aaron Nellis's 1877 patent claimed it would "keep cattle off of [fences]," as did Henry W. Putnam's patent, which declared that "animals running against or pressing upon barbed ends . . . will be pricked and turned back." John McNeill's 1877 beveled points possessed "a sufficiently offensive character for cattle," which would "avoid the danger of injuring them by penetrating too far."[51]

The recurring language about physical contact with animals, about respecting and preventing and deterring and pricking, is relatively mild. These early patentees clearly recognized the problems involved if they asserted that the new fence was effective because it was dangerous. Only one of these early patents specifically included the word "pain." Parker Wineman's 1876 pyramidal form provided great rigidity and strength "so that the points of the device cannot be bent or turned aside by the pressure of the animal's body, but will be sure to penetrate the skin and give pain, thereby causing the animal to keep at a safe distance from the fence thereafter."[52]

Avoiding the Language of Pain

By the late 1870s a second set of patents appeared in which language that included even mild references to control of animals via pricking disappeared, a clear response to the national debates about the harmful effects of barbed fencing and to the anti–barbed wire groups and state legislatures that began to oppose or restrict the use of barbed fences. These barbed fence inventors emphasized the inexpensive character, the physical integrity, and the ease of application of their inventions.

Spencer St. John's 1878 patent contained no mention of contact or injury, or for that matter, of animals or humans. Instead, the language focused on the cheapness of his improved barb: "Less material is used . . . and the advantage of less cost in manufacture is thereby gained." Montraville Mighell's 1878 patent purportedly combined cheapness and physical integrity "to produce a barbed wire which may be produced by machinery cheaply and rapidly, and one from which the barbs are not liable to become detached or loose." Issued the same day as Mighell's patent, William Warden's patent for improvement in barbed fences detailed its novelty in terms of portability, strength, and cheapness: "a fence which shall . . . possess the important characteristics of strength, lightness, durability, and cheapness, and which shall have as essential qualities convenient and thorough portability."[53] Whereas Warden included control of livestock as the purpose of his fencing improvement, the next four patents issued for improvements to barbed fencing made no references to control or to objects of control at all. Instead, the patent language concentrated on descriptions of barbs, the immobility of barbs, the lack of rust or corrosion, ease of application, and low cost.[54] By this time in the history of barbed wire, the purpose and meanings of the new fence could be understood without elaboration.

The Warning Device

Even as the patent language shifted its focus from the ability to repel animals (and humans) to the integrity and quality of the new fence, a third innovation appeared in patents: the warning device. An early reference to a "visible" fence appeared in Lucien Smith's 1867 patent for a wire fence. Smith's fence was strung with cast-iron spools with "four short wire spurs . . . projecting radially from [the spools]." Smith claimed that his fence would "turn animals readily, as they can see it better than the ordinary wire fence, which has nothing attached to the wires to attract attention, and the animals will not encounter the spurs or the spools."[55] Another early example of a warning device appeared in 1868 in Michael Kelly's "thorny fence" patent. Kelly recognized that cattle unaccustomed to barbed wire would not easily see the slender wire and small "thorns" of his fence. His patent suggested two ways to aid unsuspecting animals: (1) stringing a tarred rope along the fence, which would "perform an important function in aiding the sense of sight and of smell in detecting the presence of the fence"; and (2) forming a stone, earth, or turf ridge below the fence to "indicate to animals the locality of the fence, and thus avoid their unconsciously running against it."[56] The warning systems Smith and Kelly described led to actual patents at the end of the 1870s when public complaints against barbed wire were threatening the new fence.

One of the earliest patented warning devices was Edward Crandal's 1879 "indicator," which consisted of a "flat piece of water-proof paper, wool, veneer, or other water-proof flexible material" that was affixed to the top wire. This light and flexible material did not add weight to the fence (which would cause sagging), nor did it "tempt depredators to remove it from the fence, as in the case of the wooden blocks which make excellent kindling wood."[57] Hiram Scutt's 1880 warning device performed double duty, as barbs were added to his warning blocks.[58] In 1881, John La Fleur patented a "simple, cheap, and convenient means for making barbed wire fences conspicuous," with his tin or metal "tag" that swung freely in the wind.[59] In the 1880s alone, nearly two dozen patents were issued for some kind of warning device to be used in conjunction with barbed wire fencing. This trend, coupled with the disappearance of language about repelling animals through injury, manifested a kinder and gentler phase of barbed fencing in response to the public demand for manufacturer accountability.

Diminishing the Danger

As barbed wire manufacturers continued their campaign against detractors, and as barbed wire came into wider use despite its drawbacks, inventors took a new tack. Highlighting the harmful nature of previous barbed fencing, they claimed that their new and improved barbed wires were designed to protect animals from physical injury.[60] These improvements shared two specific elements: the ability of barbs to rotate in place and the design of the barb.

One of the earliest references to the prevention of injury is found in Thomas Orwig's 1878 patent that had barbs revolving on the wire, even as they stayed in place laterally, so as "to prevent . . . tearing the flesh of animals, as frequently occurs in the use of rigidly-fixed barbs."[61] Ironically, a patent granted the same day as Orwig's touted rigidity as one of its objectives: "[The barb] cannot be displaced, nor the points pushed back through between the strands, and thereby be rendered loose and comparatively useless."[62] It would take several more years before Orwig's approach to barbed fencing replaced the objectives of pricking, penetrating, and causing pain, but when it did, it set off a two-decade-long quest to modify the basic design in the service of safety.

William Butler's 1881 fence-barb patent, for instance, reported that "domestic animals are apt to get themselves cruelly lacerated by coming violently against the barbs of ordinary wire fences. . . . My invention seeks to overcome this difficulty" (fig. 11).[63] Butler's barb was set on a flexible spring that bent in the direction the animal moved along the fence, causing no harm to the animal and allowing the barb to spring back to its original placement. John May's 1882 patent claimed that he had created a barb that "adapted" to animal movement and thus prevented "the severe laceration and pricking of

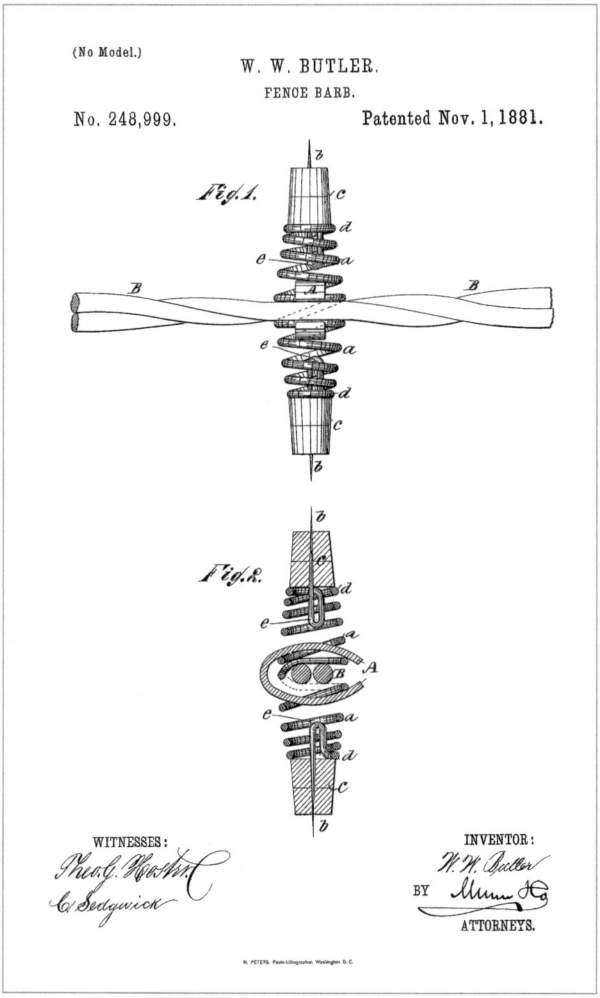

Figure 11. W. W. Butler's fence-barb patent, November 1, 1881. From US Patent and Trademark Office.

animals which result from the use of ordinary and common barbs." May's barb purportedly warned but did not injure animals: "the effect being rather to exert a scraping tendency and pressure against him than to cause a puncture or laceration." George Beerbower's 1885 patent was also designed to "provide a barb . . . that will not mutilate or injure animals." This particular barb "inclined" in either direction parallel to the fence wire so that "animals come in contact with them, and prevent lacerating their flesh, as a rigid barb will do." When Edward Robinson applied for his patent for an improved barb, he included explicit language about the need for such an invention: "Great objection has been made to barbed-wire fences because the barbs, being rigid, tear the hides of the cattle and, on the line of the highways and about the houses, the clothing of persons who perchance go near them." Robinson's patent was essentially a coiled spring with sharp ends protruding outward from a twisted wire. Although different in shape, it was similar to Butler's version of a harmless barb. Constant Trevitt and Solomon Mouck's 1889 patented barb "freely rotated" on its axis when pushed by an animal's side, "resulting in giving animals all necessary warning without unnecessary injury."[64]

Besides these rotating and inclining barbs, patents as early as 1878 began to include design changes to the standard barb. Jay Fisher claimed his "prongs" were "constructed differently from the barbs heretofore used. . . . Instead of having sharp points, which lacerate the animals which come in contact with them, the projecting ends are made angular or blunt, so that they will not readily penetrate the skin or flesh of the animals, but will, through contact, sufficiently warn them of their proximity to the fence." The language of Fisher's patent emphasized the danger, pointing out that the use of "sharp-pointed barbs has proved injurious through the wounds they inflict on the animals that come in contact with them. The object of my improvements is to provide a fence which possesses all the advantages of a wire fence without having the objectionable features . . . one which can be seen by animals, and . . . will not injure them."[65]

When the rotating and altered barbs failed to attract commercial attention, inventors employed the rhetoric of safety and proposed barbless barbs. Hugh Gearty's 1892 patent described his attempt "to obtain a safety-barb that will be sufficiently severe to make the fence stock-proof and that will not greatly injure animals if they are forced over the wires." Gearty's barb revolved perpendicular to the wire so that "animals will not have their flesh torn or be severely injured by falling on the wire or by dragging over it. It is for that reason I term it a 'safety barb.'"[66] Not to be outdone, Fitz Longstreet's 1902 patent, designed to repel but not injure animals, demonstrated the extreme direction inventors moved over time in response to the harmful nature of barbed fencing. Longstreet's barbs were not sharp and prickly, but in fact their ends were "rounded or ball-tipped," reminding twenty-first-century observers of jacks (from a ball and jacks game) that seem like harmless fun strung along a steel wire. What effect this particular barb had on animals can only be imagined, although animals must have been disgusted at the worthlessness of a fence not even useful for scratching.[67]

These developments overlap to some extent, but a pattern emerges that makes sense in the contexts of pressing need, celebrated discovery, concerned backlash, and expanding usage. Patents of the late 1860s to mid-1870s added barbs to smooth-wire fencing to control stock and to protect fences from destruction by unruly animals. That this barbed wire exploited its ability to cause pain was an unacceptable outcome for many, and so warning devices were patented. By the late 1870s, the language used in patents ignored possible harm in the face of public backlash and focused on the physical integrity of the fence. Multiple patents then appeared that emphasized the damage caused by barbed wire and promised to overcome it.[68] In the last decade of the nineteenth century, only 59 patents were issued for barbed fencing, a substantial drop from the 238 patents of the previous decade. In the end, the original Glidden and Haish patents held sway. None of the hundreds of so-called improvements proved to be practical. To be effective, a barbed wire fence had to inflict pain. And to be cost-efficient, it had to remain simple.

Washburn & Moen's concerted effort to defeat state legislative efforts to ban or regulate the use of barbed wire was largely successful. It is not clear why barbed wire opponents testifying before legislative bodies did not draw on the creation of the new curative industry, the escalating damage done to livestock, or to the lightning-caused electrocutions via barbed wire as evidence of the cruel nature of barbed fencing. What is clear is that the Washburn & Moen Company seldom had to respond to actual contemporary evidence that would have confirmed the actual, not alleged, vicious nature of the "Devil's Rope."

Besides defending their products, barbed wire manufacturers turned to a relatively new medium to sell their barbed wire: broadscale advertising. The images and verbiage with which barbed wire was sold in the final quarter of the nineteenth century tell a very interesting story.

CHAPTER THREE

"The Perfect Fence"

Selling Barbed Wire

What do advertisements mean? . . . They urge people to buy goods,
but they also signify a certain vision of the good life;
they validate a way of being in the world. They focus private fantasy;
they sanction or subvert existing structures
of economic and political power.
— Jackson Lears, *Fables of Abundance*

Barbed wire was invented, as Reviel Netz writes, "to prevent the motion of cows; its function relied on violence; its success depended on deployment on a vast scale."[1] Before barbed wire could be widely deployed, however, before the owners of valued livestock would entrust their animals to a dangerous fence, a vigorous backlash against its violence had to be addressed. Barbed wire inventors worked to please potential customers by shortening and modifying the barbs and by devising warning strips. Published appeals to legislatures and to the general public were effective to some extent. But finally, to sell their wire, manufacturers turned to advertising and did what advertisers have always done: they played on the fears and beliefs and desires of potential buyers. Specifically, besides touting their product as the answer to the pressing fencing question, barbed wire advertisers argued their case with anthropomorphized animals; celebrated their companies' technological and industrial capabilities; promoted their fences' ability to exert control by playing on racial, class, ethnic, and gender prejudices; employed unabashedly religious imagery; and invoked a comforting, protective, and patriarchal domesticity. The visual and verbal images with which barbed wire was sold in the final quarter of the nineteenth century tell a very interesting story about how manufacturers wanted their fences to be perceived, how they worked to construct those perceptions, and, indirectly, what kinds of Americans bought and used it.

The question of the relationship between advertisements and the people they are meant to influence is not entirely straightforward. Richard Ohmann argues that as advertisers constructed the meanings of their products, they were simultaneously constructing the lives and social relationships of their customers. In this view, advertising both constructed and mirrored consumer desires and beliefs.[2] Pamela Walker Laird disagrees, claiming that we have little access to information on consumer responses and must withhold judgment.[3] While Laird's caution is important to bear in mind, we will assume with Ohmann that barbed wire advertisers at least somewhat accurately assessed attitudes of the customers who bought the new fence so avidly and that the advertising in turn influenced the identities of consumers.

Barbed wire advertising in the last decades of the nineteenth century took place in the context of newly opened national markets, of newly available manufactured goods, and of the new modes of advertising that made nationwide sales of the products possible. "Between 1870 and 1900," Robert Jay writes, "the volume of American advertising increased by more than tenfold."[4] In the sixty years between 1860 and 1920 advertising became a coherent industry in the United States, involving a wide-scale corporatization that brought advertising into most Americans' daily lives.[5]

Along with the increased volume came new methods of advertising. In 1855 the Russian populist Aleksandr Herzen claimed that impresario P. T. Barnum "understood the great secret of the age of rhetoric, the age of effects and phrases, exhibitions and advertisements; he comprehended that the most important thing for the contemporary nominalist is the poster!"[6] After Barnum's success, Laird writes, "'to bill it like a circus' became a common expression among ambitious advertisers."[7] In terms that are true for barbed wire advertisements as well, Cary Nelson and Michael Chasar note that patent medicines and soaps

> tested market strategies, experimented with multiple print media ranging from farmer's almanacs to fliers and trade cards, and solicited consumer involvement and input in the marketing process with contests and entertaining rhetoric. . . . Filled with testimonials, poems, moral warnings, and medical advice, such almanacs were omnipresent. Families consulted them throughout the year, and, unlike a newspaper issued by a publisher, a patent medicine company could control the publication's entire contents, effectively creating an entire magazine

devoted to a single product. Like advertising trade cards, which began to be popular in the 1870s and became ubiquitous in the 1880s and 1890s, almanacs were entirely under the company's control.[8]

Haish, Ellwood, and Washburn & Moen made creative use of trade manuals, almanacs, fliers, posters, and trade cards to define their fences and promote their distinctive brands. The meanings constructed by these various kinds of barbed wire advertisements are the focus of this chapter.

Trade Manuals

Washburn & Moen and Ellwood's 1876 *Barb Fence: Its Utility, Efficiency and Economy* was an early trade publication meant to introduce a product that was beginning to revolutionize the fencing industry. The contents of the forty-two-page booklet provide a good introduction to the advertising strategies that follow (fig. 12).

Before the verbal argument commences, images of Washburn & Moen's Grove Street Works and the Quinsigamond Works illustrate the company's industrial capacity. Atop the Grove Street Works' imposing structures, a waving American flag and smoke rising from nine smokestacks supply kinetic proof of the demand for barbed wire and of the company's ability to meet it. The Quinsigamond Works, slightly smaller in scale, still boasts eight belching smokestacks.

In text framed on each page by a decorative strand of barbed wire, the manufacturer opens its case for adoption of the "barb fence" with a self-serving truism: "It is characteristic of great and useful inventions that their simplicity and perfect adaptation to the desired end at once suggest the question, 'Why was not that thought of before?'" For a quarter of a century, plain wire has been the answer to the fencing question, the booklet points out, "especially in the sparsely wooded regions of the West," and Washburn & Moen was the first company to offer plain wire for fencing in the West. A single strand of plain wire, however, is easily broken, cannot be seen, is adversely affected by changing temperatures, and provides no opposition to animals that push against it and break through. These defects have been obvious for years, the account continues, "and many have been the attempts by thoughtful persons, during that time, to overcome its defects, and to raise, by some new device, this strong, cheap form of fence to the dignity of something fully answering

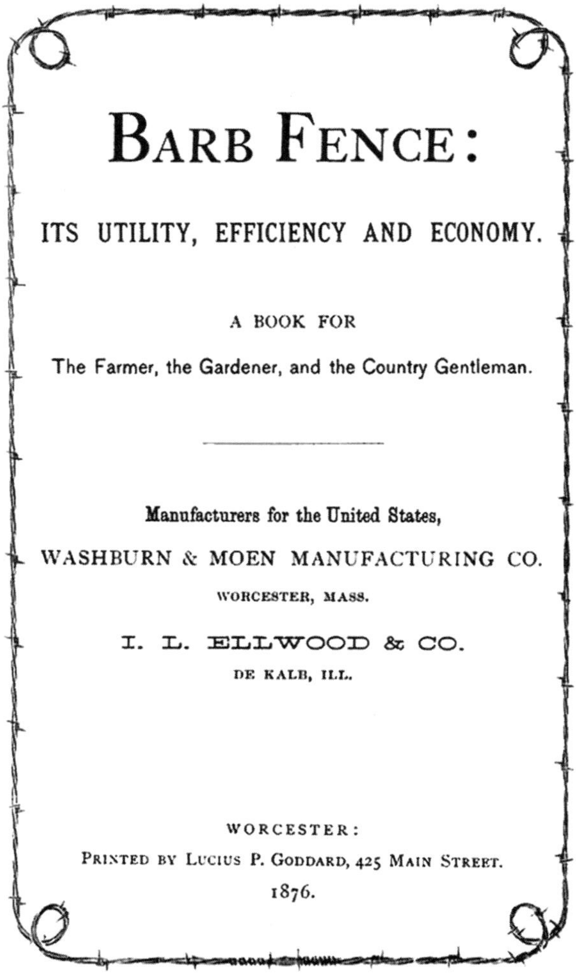

BARB FENCE:

ITS UTILITY, EFFICIENCY AND ECONOMY.

A BOOK FOR

The Farmer, the Gardener, and the Country Gentleman.

Manufacturers for the United States,

WASHBURN & MOEN MANUFACTURING CO.

WORCESTER, MASS.

I. L. ELLWOOD & CO.

DE KALB, ILL.

WORCESTER:

PRINTED BY LUCIUS P. GODDARD, 425 MAIN STREET.

1876.

Figure 12. Title page of Washburn & Moen's and I. L. Ellwood's 1876 booklet *Barb Fence: Its Utility, Efficiency and Economy.*

the purpose." The fence with such dignity, of course, is the Glidden *Steel Barb Fence*, which "accomplish[es] the long sought desideratum."[9] High-flown rhetoric for a steel fence!

The caption in figure 13 (reproduced in several Washburn & Moen publications and as a trade card) asserts that "BARB FENCE protects the most tempting Crops from the most unruly Cattle." The obviously frustrated cattle trot away or investigate the fence while the abundant and perfectly protected crop flourishes on the other side of the fence that so decorously orders the scene. A

Figure 13. "Barb Fence protects the most tempting Crops from the most unruly Cattle." From *Barb Fence: Its Utility, Efficiency and Economy*, 1876.

substantial farmhouse stands in the rear, promising prosperity for wise users of the fence. Three pages later a second image depicts an orchard protected from ravenous cows by barbed wire. The fence borders a road on which seven men in suits and hats stand in conversation. An even larger house than the one in the first image rises above the scene. Prosperity through order established by barbed wire is the message.

After a section laying out the advantages of barbed wire over plain wire (two strands of steel cannot be broken nor are they adversely affected by temperature changes, the barbs make it more visible, and cattle that come in contact with it avoid it thereafter), the booklet explains how barbed wire can serve the genteel clientele it addresses—the farmer, the gardener, and the country gentleman:

> The *Farmer* can perfectly protect his most tempting crops from the most unruly cattle.
>
> The *Professional* man or *Mechanic*, living in villages, can throw around his vegetable, fruit or flower garden, a barrier absolutely impassible by man, child, beast or animal.
>
> The *Connoisseur* in any of the choice productions of nature, and living in the suburbs of any town or village, can, by the adoption of the Barb Fence, be sure of the safety of his favorites from any foe from without.

The Barb Fence is the only means for obliging *every body* and *every thing* to enter a man's premises by the same way that he himself enters.

The adoption of the Barb Fence is the only method by which an owner can be certain that no *man* or *beast* is on his premises without his wish or knowledge.[10]

All five of these assertions are about protection: protection of crops, gardens, and property. Men, women, children, beasts, animals, unruly cattle, and *every body* and *every thing* (just to make sure) are the foes that the impassable barrier will oblige to stay outside. Although the testimonies that appear later in the booklet refer to the fence's ability to control the fence builder's own animals, barbed wire at this point is a tool to keep intruders out, whether human or animal.

Emphasizing the human element, the booklet also presents barb fence armor, a device consisting of "convolved circles of Steel Barb Fencing fastened upon strips of wood twelve feet long, which can be secured upon the upper edge of any wooden fence, making by the combination: First. A fence absolutely impassible by man or beast. Second. A fence no person can sit upon" (fig. 14).[11] An etching depicts a paying crowd of spectators separated from the nonpaying crowd by a fence topped with barbed wire. Near the end of the booklet, N. F. Apolonio, president of the Boston Base Ball Association, testifies that five hundred feet of the armor "saved us, the afternoon of the first day it was put on, fifty dollars ($50) in special police, and from $400 to $500 admission fees."[12]

After an examination of seven patent claims, important largely because of the controversy already brewing with Jacob Haish, the booklet offers sixteen pages of testimonials "sent us spontaneously from different parts of the country, where the Barb Fence has been introduced."[13] Only two years after barbed wire was first manufactured, these testimonials from Illinois to California and from Michigan to Texas indicate how far the new wire has already spread and give a sense for how it has been received. A few examples illustrate the enthusiasm and lay out various claims for the fence.

Moline Plow Works, Office of Deere & Co. Moline, Ills.,
June 15th, 1875

Dear Sir: In reply to your letter of inquiry in regard to the Barb Wire Fence, I would say that I find it a complete success, as no stock will go through it,

Figure 14. "Barb Fence Armor.—A Ball Ground enclosed with 'Barb Armed' Fence." From *Barb Fence: Its Utility, Efficiency and Economy*, 1876.

and I consider it the best and cheapest fence in use. I am waiting for that last ordered, and shall use a large quantity after harvest.

Truly yours,
JOHN DEERE[14]

Yes, that John Deere. That he replied "to your letter of inquiry" suggests that the testimonials may not have been as spontaneous as claimed.[15]

Panora, [Iowa], January 25th, 1876

Gents: Having handled your Glidden Patent Barb Wire the past season, we take pleasure in recommending the same. We have tried several kinds of barbs and barb wire, and pronounce the Glidden Barb Wire the greatest invention of the day.

Yours very respectfully,
JACKSON & GARLOW[16]

Merritt Island, Tolo [*sic*] Co. [California], April 17th, 1876

Gents: In reply to your note, I am prepared to say that having given the Glidden Barbed Wire Fence a fair trial, I am satisfied that it is the most effectual fence against cattle and horses ever built; and for cheapness, facility with which it can be put up, and apparent durability, it is superior to all other fences. I regard it as one of the great inventions of the age, and in the near future it will revolutionize the whole fence question in the interest of the farmer.

Yours truly,
J. B. Greene[17]

One of the great inventions of the age, revolutionary, most effectual, superior—the enthusiastic J. B. Greene might profitably have given up farming for a career in advertising. Or, and this is more likely, he is simply parroting the manufacturers' own rhetoric. Other testimonials follow suit.

Stockton, Cal., April 11th, 1876

Dear Sirs: Our sales of "Glidden's Patent Steel Barb Fence Wire" . . . have been very satisfactory. . . . It makes the cheapest and most durable fence in known use, standing wind, flood and fire, and is perfectly stock proof.

Respectfully,
H. C. SHAW PLOW CO.[18]

Plano, Tulan Co., Cal., April 19th, 1876

Dear Sirs:

My field was surrounded by large bands of wild cattle and horses, which broke through the plain wire repeatedly, but when I constructed a strong fence in place of the plain wire, they have failed to get into the field. I consider it one-third the cheapest and much the best material for fencing I have ever seen. It is a terror to horses and cattle. As soon as practicable I intend to build several miles of this fence.

Yours truly,
S. A. BARKER[19]

S. A. Barker's unabashed use of the word "terror" in a letter published by the manufacturer indicates that serious backlash against the dangerous barbs had yet to surface.

Malta Ills., January 22, 1876

Gents: During the past twelve years I have been in the hardware and farm implement trade. I have never introduced an article of any kind that gives such universal satisfaction as your Barb Wire Fence. I have yet to hear the first complaint. All things considered, it is the best and cheapest fence in the market, or in use. Have sold about 7,500 pounds last year, and I expect a large increase in sales for next year. It is a perfect antidote for breachy stock.

Yours respectfully,
GEO. W. SMILEY[20]

Lone Pine, Ingo Co., Cal., April 17th, 1876

Gents: In answer to yours, we can say that we are well pleased with the Glidden Steel Barbed Wire. Two strands of wire on a ditch effectually turns breachy cattle that have been bothering for eight or ten years. Now we can go to bed at night and rest contented about our crops, and our stock not being stolen. The Indians have bothered us a great deal, in cutting our fences, and going through on horseback, letting our stock out, and outside stock in the ranch. Since we put up the wire they have not bothered any. Do not think it necessary to send certificates from this county, as the barbed wire is well known. It is cheaper fencing by one-half than any other. Think it is the greatest timber saving invention of the age.

Yours very respectfully,
C. W. Johnson, A. H. Johnson[21]

By this account, the Glidden fence worked equally well against strong-willed or breachy cattle and trespassing Indians. Native Americans appear here, as they will in later advertising, as undesirables who are best fenced out. We return to that idea later.

After the testimonials and following eight pages delineating even more patents controlled by Washburn & Moen, "Notice is hereby given to Manufacturers of Barbs and Barbed Fence Wire, and those purchasing, selling and using Barbs and Barbed Fence Wire, which infringe any of the foregoing claims . . . [that] both the seller, purchaser and user will all be held responsible."[22] Less than two years after the company commenced manufacturing barbed wire, Washburn & Moen asserts its control over the industry that will soon approach monopoly status.

In the 1878 edition of this publication, the testimonial section expands to twenty-nine pages and includes press reports. Most of the comments repeat the sentiments of those previously quoted, but there is an occasional gem, like

Stephen & Prestegad's hyperbolic claim from Lee, Illinois, that barbed wire "will stop anything in the shape of stock, from a grasshopper to an elephant."[23] A statement taken from the *Boston Journal of Commerce* boasts that barbed wire, "while four years ago as little known as was the telephone in the time of King Pharaoh, is to-day actually in use to the extent in weight of eight thousand tons, or more than fifty thousand miles in length!"[24]

In subsequent years, Ellwood and Washburn & Moen jointly published several trade manuals meant to expand public knowledge of fencing and of their barbed wire. These include an 1879 publication running to 118 pages and with an equally imposing title: *The Fence: A Compilation of Facts, Figures and Opinions from National and State Agricultural Reports, Agricultural Journals, and the Public Press, for the Past Sixty Years, (Presented in Their Original Form,) Showing the Waste and Need of Fencing Material, the Cost of Fences, Their Waste and Wear in the United States, Relation to Other Farm Property, etc., etc. and a Compilation of Statutory and Judicial Facts and Decisions in the State and Federal Courts, Relating to Railway Fences and Fence Laws, with the Early History of Enclosures*. The manufacturers lay a historical foundation for their product with topics that include fence laws in various states, herd laws, values of property, and railroad fences. Praising the English Enclosure Acts, the publication quotes A. Fitzherbert's *Book of Husbandry* to the effect that "if an acre of land be worth Sixpence before it is enclosed, it will be worth Eightpence when it is enclosed."[25] A "No-Fence Theory," according to which "the highest degree of rural beauty is a village without fences or other distinctive marks to properties," is rebutted by the claim that "*the love of exclusive possession is the mainstay of society*" and "well defined boundary lines to property greatly enhance its enjoyment."[26]

The following year Washburn & Moen published a twenty-eight-page booklet titled *Steel Barb Fencing, the World's Fence, Everywhere Adapted to All the Uses of Husbandry in All Climates and under All Exposures, "the Perfect Fence."* This publication includes images of Brazilian and Australian formal gardens in which barbed wire is in use and lays out the advantages of barbed wire over other kinds of fences.[27] And in 1882 Washburn & Moen published a forty-seven-page booklet titled *The Fence Problem in the United States, as Related to General Husbandry and Sheep Raising: Facts and Statistics from Authoritative Sources with a View of Fence Laws and Customs, Illustrated.* Most interestingly, *The Fence Problem* lays out a history of exponential growth of the industry: from 10,000 pounds of barbed wire made and sold in 1874 to 120 million pounds in 1881.[28]

These and other trade manuals were supplemented by an advertising medium meant to inform and influence the public in a less formal mode: the almanac. Jacob Haish's *Barb Fence Regulator* was the first to appear, and Ellwood's *The Glidden Barb-Fence Journal* followed in short order.[29] The stark contrast between the competitors' publications is fascinating, with Haish's work personal and oppositional and Ellwood's publications bearing a more corporate stamp.

Almanacs

As underdog in the barbed wire patent, manufacturing, and sales wars, Haish employed what he described as new tactics "of a guerrilla order, very spectacular in many respects, yet catchy in their make-up."[30] Most prominent in Haish's guerrilla arsenal was the *Barb Fence Regulator*, an almanac that grew from eight pages to sixteen and appeared more or less annually for more than a decade. Numbers printed in successive issues proclaim that by 1887 five hundred thousand copies of the *Regulator* had been disseminated (distributed largely by wholesalers who stamped their company information on the back covers).[31] Haish boasted about the role his *Regulator* played in the advertising revolution: "Nothing like it had ever before appeared in print, quotations from celebrated authors, caricatures and cartoons of the most grotesque character, poetry . . . appealing to the sensibilities and the risibilities of the general public." Haish also claimed that "while my opponents knew well how to take care of the legal end, they were entirely at sea when they sought to catch public opinion, a new era in advertising had dawned which carried them off their feet."[32] Just what were the sensibilities and risibilities to which Haish appealed? The name *Regulator* provides one indication. It likely refers to the extrajudicial group by that name that carried out lynchings to rid several Illinois counties of notorious horse thieves, murderers, and arsonists in 1841.[33] Haish saw the battle against his monopolist rivals in exactly these regulatory terms, although his work was done with words and images rather than with a noose.

Haish's almanac employed humor that varied between sophisticated and bawdy, playing on the diverse backgrounds of potential barbed wire consumers. In volume 2 of the *Regulator*, for example, "Darwin now thinks he is on the right track for that missing link. He is studying the play of Hamlet, and is paying particular attention to the Ghost of Hamlet's father, as that personage 'could a tail unfold.'"[34] The humorous stories often involve

Haish's "S" Barb and/or the monopolists' barbed wire, as in this interpretation of a nightmare: "If you dream of a ring-tailed monkey chewing a bar of Loomis Bros. bar soap on the top of an 'S' Barb Steel Fence Wire, and scratching his back with the MONOPOLY BARB, it means that when you go hunting on the Mississippi River you will have fisherman's luck—'wet feet and a hungry gut.'"[35] The *Regulator* can be slightly ribald, as in an image that shows a salesman for the monopolists' barbed wire hitting himself in the head as two buxom girls come down a set of stairs. The text accompanying the picture tells the sad tale in German-inflected English, as so often in Haish's publications:

> Der Monopoly Barb Fence Drummer's Mistake. The farmer told the Drummer that it would be necessary for him to sleep with the children or with the farmer himself, as their accommodations were limited. Quick as lightning the Dutchman expressed his resolution not to sleep with the "boddoration shiltren," so he slept with the farmer. The rest of the story may be given in his own language. Vell, in de mornin', ven I comes mit der sthairs down, I see two girls apout seventeen unt nineteen years old, unt I ask der old man: 'Pees dem girls der shiltron you told me upout?' unt he say 'yaw: dem ish mine only shiltren!' unt I say to myself, "Py shiminy! Ish dot so!"[36]

Poetry was another vehicle for Haish's antimonopolist message. According to Nelson and Chasar,

> If you wished to give your woefully material product aspirations to a higher life, if you wished to supplement the apparently mundane with whimsical pleasure, if you wished to promote sales by appealing to your buyer's potentially broader cultural inclinations, if you wished to create a readerly community bemused by rhetorical excess yet curious about the product nonetheless, then for well more than half a century you hired a poet.[37]

The *Regulator* is rich with maudlin and sometimes witty poetry, most commonly bowdlerized Shakespeare. In a version of Julius Caesar's speech to Marc Antony, Haish inserts his rival Washburn into a damning context: "Let me have men about me that are fat; / Sleek-headed men, and such as sleep o'nights; / Yond' *Washburn* has a lean and hungry look; / He thinks too much;

such men are dangerous."[38] "Dangerous men" of the monopolist strain are mocked mercilessly. "A Texas Idyl" in volume 6 of the *Regulator*, for instance, begins with lines from Charles Wolfe's well-known "The Burial of Sir John Moore after Corunna," but instead of honoring the British hero killed and buried with simple honors as does the original, the *Regulator* version ends with a Texas surprise:

Not a drum was heard, not a funeral note,
As his corpse in the woods we hurried;
Not a ranger discharged his farewell shot,
O'er the grave where the stranger was buried.
They buried him darkly, at dead of night,
Without any pomp or pageant;
As they shoveled him in they said "*all right*,"
He's a Monopoly Barb Fence Agent.[39]

On the same page there is a parody of Edgar Allan Poe's "The Raven" that ends with three monopolists deciding there's no way to best Jacob Haish. There are, in fact, so many poems in the successive volumes of the almanac that a reader may be forgiven for seeing the poems as barbs strung along the twisted wires of Haish's multimedia appeal to the general public.

The *Regulator* was informative as well as entertaining. It included recipes, advice on when to cut timber, suggestions for protecting farm implements against rust, recommendations on what to do with a fainting person, and thoughts on how to soak woolen fabrics to keep them from shrinking—all standard material for an almanac. A reader moves back and forth from these bits of practical wisdom to explanations of the various benefits of the "S" Barb. A page in volume 2, for instance, declares at some length why Haish's fence is "the Best, Cheapest and Most Durable," gives directions for putting up barbed wire, states the price at the factory in contrast to the costs of other types of fencing, and reminds the reader that *Regulator*s are "sent free to any address upon application."[40]

Although the early *Regulator*s both entertain and inform, they also are vehicles to decry Ellwood and Washburn & Moen as malicious and un-American monopolists. As noted previously, they make their case in the context of widespread antimonopoly sentiments at the end of the nineteenth century. Jacob Haish would not ultimately beat the barbed wire monopolists, but he caused them considerable trouble as he ginned up public opinion against them.

In volume 2 of the *Regulator*, for instance, bold headlines in various fonts proclaim:

BARB FENCE WIRE MONOPOLISTS! CORNERED AT LAST
DEATH KNELL
UNCONDITIONAL SURRENDER
OH, YE GODS!
UPRIGHT AND JUST
WHOSE TURN IS IT TO EAT CROW[41]

The uninhibited populist tells his readers that "some of the best legal talent in the West who are disinterested in the result and were not employed to express an opinion, have been permitted to look at and examine the testimony and proofs, and their unanimous verdict has been that the 'Broad Claim' cannot be sustained and is as worthless as a *tin tit*." *The Regulator* depicts the monopolists and their "perfect fence" as the devil incarnate, a cloven-footed, horned, sharp-tailed devil with the "broad claim" to all the barbed wire patents sticking up out of his back pocket (fig. 15).[42] "ANCIENT ANCESTRY," the headlines shout: "THE FATHER OF LIES. THE MONOPOLISTS' PEDIGREE. THE BOSS BULL-DOZER." Used in many contexts in Haish's advertising, the term "Bull-Dozer" first appeared during the contested election of 1876 and referred to those who resorted to intimidation and violence against potential voters. Haish's repeated use of the term in reference to Washburn & Moen aligns that company with similar methods of intimidation and coercion.[43]

A full-page cut of a field fenced with Haish's "S" Barb wire features three men in suits under the weight of a yoke labeled "broad claim" and pulling a wagon driven by Haish's familiar portly German American and his ample wife (fig. 16). The man wields a long whip and holds reins connected to rings in the noses of three monopolists. The woman holds up a flag that reads "We aim to teach monopolists that the American People will not submit to extortion nor undue oppression." At the bottom of the page stands the headline PROPHECY FULFILLED, followed by a curse from Isaiah: "for we have made lies our refuge, and under falsehood have we hid ourselves."[44] Populism, religious indignation, and pictorial humiliation—Haish does not miss a trick.

Figure 15 (opposite). Jacob Haish's depiction of his antagonists as monopolist devils in the legal fight over patent rights. From the *Barb Fence Regulator*, 1877(?).

A. A.

ANCIENT ANCESTRY.

THE FATHER OF LIES.

THE MONOPOLISTS' PEDIGREE.

THE BOSS BULL-DOZER.

Look!!!! Where the Broad Claim has Gone To!

MR. CLOVEN FOOT'S CARD.

When you want---STILES---give me a call. I can furnish all the latest fashions in "Monopoly," "Bull-Dozing" and "Falsifying," from a "Washburn Print" to the low grade of "Millington" Shirtings. And can warrant a perfect fit for every---suit---at law, as I have a perfect system for cutting--Lies--Blackmail, Duplicity and Fraud out of new cloth, thoroughly sponged in Hypocrisy and Deceit.

You are cautioned against using my celebrated "Trade Mark," the "Bevel Cut," or you will be sued for "Infringement."

APOSTROPHE.

"Glorious wert thou, Athens,
In thy classic grace."
But far more glorious,
Beauteous, art thou, oh
Son of *night!*
With thy artistic and classic
Appendage, *the Bevel Cut.*

Æsops Fables.

The lion, the fox and the ass entered into an agreement to assist each other in trade. Having received a large booty, the lion, on their return from the west, asked the ass to allot his due portion to each of the three partners in the treaty. The ass carefully divided the spoil into three equal shares, and modestly asked the two others to make the first choice. The lion, bursting into a great rage, denounced the ass. Then he requested the fox to do him the favor to make a division. The fox accumulated all that he had made into one large heap, and left to himself the smallest possible morsel. The lion said, who has taught you, my very excellent fellow, the art of division? He replied: I learned it from the ass by witnessing his fate.

Moral: Happy the man who escapes the toils of the monopolists.

Verified by combination entered into December last, 1878.

—o—

A lion, greatly desirous to capture a bull, and yet afraid to attack him on account of his great size, resorted to a trick to ensure his destruction. He approached him and said, "I have slain a fine sheep, my friend, and if you will come home and partake of him with me, I shall be delighted with your company." The lion said this in the hope that, as the bull was in the act of reclining to eat, he might attack him to advantage, and make his meal on him. The bull, however, on his approach to his den, saw the huge spits and giant caldrons, and no sign whatever of the sheep, and without saying a word, quietly took his departure. The lion inquired why he went off so abruptly, without a word of salutation to his host who had not given him any cause of offence? "I have reasons enough," said the bull." "I see no indication whatever of your having slaughtered a sheep, while I do see, very plainly, every preparation for your dining on a bull.

Haish's interview with Washburn.

—o—

An ass having put on the lion's skin, roamed about in the west, and amused himself by frightening all the foolish manufacturers of barb wire he met within his wanderings. At last, meeting a fox he tried to frighten him also, but the fox no sooner heard the sound of his voice, than he exclaimed, "I might possibly have been frightened myself if I had not heard you bray."

Make your own application.

MONOPOLISTS HAVING TAKEN THE YOKE MUST WEAR IT.---PROPHECY FULFILLED.

Oh! then our best blood turn
To an infected jelly, our necks yoked with the
Broad Claim, and our names
Be yoked with his that did betray the best.
Turn then our freshest reputation to
A savor, that may strike the dullest nostril
Where we arrive, and our approach be shunn'd,
Nay, hated too, worse than the greatest infection
That e'er was heard or read.—*Shakespeare.*

"But the word of the Lord was unto them precept upon precept, line upon line, here a little and there a little, that they might go, and fall backward, and be broken. Because they had said, we have made a covenant with death, and with hell have we an agreement; when the overflowing scourge shall pass through, it shall not come unto us; for we have made lies our refuge, and under falsehood have we hid ourselves.

"Judgment also will I lay to the line, and righteousness to the plummet; and the hail shall sweep away the refuge of lies and the waters shall overflow the hiding place. And your covenant with death shall be disannulled and your agreement with hell shall not stand; when the overflowing scourge shall pass through, then shall ye be trodden down by it. From the time that it goeth forth it shall take you, for morning by morning shall it pass over, by day and by night; and it shall be a vexation only to understand the report."—*Isiah.*

Year after year the *Regulator* excoriated the monopolists. Volume 4, for instance, begins with a letter from Haish praising the faithful dealers and consumers of the "S" Barb: "men whose convictions of right and wrong, are based upon the 'eternal fitness' of things, backed by a reserve force, resting in an All-wise and truth loving God, who will bring to naught all the wicked machinations of Godless monopolists and their bull-dozing allies."[45] Despite the efforts of these upright, divinely sanctioned purveyors of his "S" Barb fence, a time has arrived, Haish warns, "when a few moneyed aristocratic monopolists can 'pull the wool' over the eyes of an intelligent public." "We are not fond," Haish continues,

> of filling your ears with language that may seem harsh and severe nor of using strong measures to sustain our principles, but we do feel justified in using every weapon of offense and defense that shall protect us and our business from such unwarranted and high-handed outrages, such as unprincipled monopolists are perpetrating from time to time, to prop up and sustain a cause rotten to the core.[46]

Volume 4 of the *Regulator* was typeset while the patent question was still alive in the courts. It was published, however, some weeks after the decision of Judge Blodgett on December 15, 1880, declared Haish's "S" Barb an infringement on the Hunt, Kelly, and Glidden patents owned by Ellwood and Washburn & Moen. The Blodgett decision is depicted in the *Regulator* both as Haish envisioned the decision and then as it actually turned out. In the first cut, the "S" Barb cock stands triumphant over the cocks that represent the patents of barbed wire inventors Glidden, Kelly, and Hunt. An altered image was added below the original cut, with the Kelly and Glidden patents triumphant and Haish's own patent dead in the right foreground. The caption to the new image proclaims victory nonetheless: "But dot 'S' Barb Steel Fence Wire is the Boss all the same" (fig. 17).[47]

Although defeated in court, Haish continued his campaign against the monopolists in subsequent volumes of the *Regulator*. Volume 6, for instance, presents Haish's "Reasons for Accepting the License," arguing that the compromise between Washburn & Moen and himself will protect his customers from lawsuits and that he too will escape liability. Having said that, the next

Figure 16 (opposite). Jacob Haish's depiction of Glidden, Ellwood, and Washburn as monopolists under the yoke of their "broad claim." From the *Barb Fence Regulator*, 1877(?).

p in

nt,

pye.

I learn some tings vot do me goot,
Dot glory vos but a pbig name,
Unt its arrow you any vay shoot,
Vill puncture dot soap bpubble fame.

So hafing done all vot I couldt,
Dot der past vill amply attest,
Let oders dake glory for foodt
Unt I vill dake quiet unt rest.

f hope

d god-

But dot "S" Barb Steel Fence Wire is the Boss all the same.

page features the image of the bearded, horned, sharp-tailed, cloven-footed devil seen in earlier volumes, excoriated here with biblical thunder: "Behold the Devil shall cast some of you into prison, that ye may be tried, and yet shall have tribulation ten days, unless you can fill his skin with barbed irons or his head with fist spears, although he spreadeth sharp pointed things upon the wire." Fulminations throughout the volume argue that "THE TRIANGULAR FENCE MONOPOLY otherwise known as the WASHEDELWOOD BURNT MOEN & CO. RING!" and the "TRIPARTITE COMBINATION" are bogus in their patent claims. Eventually the *Regulator* turns away from the battle against the monopoly, but the early volumes caused enormous trouble for Ellwood and Washburn & Moen as Haish tapped into broadly based national antimonopoly sentiments.[48] Ellwood, as noted previously, expressed relief that a settlement with Haish had been reached, for "had no settlement been made, he would have caused us a great deal more trouble than we had anticipated."[49]

Taking his cue from Haish's *Regulator*, Ellwood began publishing what he called the *Glidden Barb-Fence Journal*. Like the *Regulator*, editions of the *Journal* feature material commonly found in almanacs. Volume 2, for instance, is replete with poetry like "The Farmer's Soliloquy," a skillful parody of Hamlet's famous monologue:

> To fence, or not to fence, that is the question:
> Whether 'tis better on the farm to suffer
> With straying cattle, or the herders fees,
> . . . .
> Or, to take up arms against a sea of trouble.
> . . . .
> A plague upon the fences that erstwhile were used.
> Were it not for this new fangled thing they call Barb Wire
> Me thinks I'd go and hang myself.[50]

There are poems that are jokes ("POET-TREE: Oak, Caroline! Fir yew I pine; / O, willow, willow, will you not be mine?"), and readers are regaled with puns ("Say, John, is your sweetheart a factory girl? Yes, William, satisfactory").

Figure 17 (opposite). Initial cut shows Jacob Haish's patent cock as victorious. After the court decision, a revised cut depicts his competitors' cock as triumphant. From the *Barb Fence Regulator*, 1881.

Humorous references to barbed wire abound, some of them mirroring and/or modeling comments found in newspapers of the time: "Stern reality. The pricks caused by sitting on a barbed wire fence," and "Yes! barbed wire costs something, but the hired man don't sit on it five minutes every time he climbs the fence. It saves time."[51] Alongside the poetry and humorous asides, the *Journal* offers farming tips, claims of superiority ("The Glidden Barb has a needle point, the sharpest of any in the market"[52]), and responses to the bad press Ellwood and Washburn & Moen had been getting:

> It is amusing to witness the spasms of certain newspaper editors and others, who assume the role of the farmer's friend and protector in their flounderings over the Court Decision sustaining the barb wire patents. Many of their statements, which they assert as facts, are so wide of the truth, and so ridiculously absurd that it is a waste of words and time to deny them. . . . There have been millions of dollars saved to the farmers by the introduction of barb wire, which never would have been known or introduced except for the patent laws.[53]

Beyond the entertaining, informative, corporate, and antimonopoly aspects of the early advertising, a closer look at the *Regulator*, the *Journal*, and various trade cards and posters produced by the competing companies reveals an interesting set of shared themes. Anthropomorphized animals argue the case for barbed wire. Industry and technology are celebrated. Racial, class, ethnic, and gender prejudices are used to promote the fence's ability to exert control. Religious imagery abounds. And finally, a reassuring domesticity is invoked. In all of these attempts by manufacturers to construct the meanings of barbed wire, nineteenth-century mores and prejudices are on display.

Anthropomorphized Animals

In *Fables of Abundance*, Lears notes the abundant use of animals in advertising during the mid- to late nineteenth century: "Advertising was . . . full of humanized animals: cows in spectacles, shawls, and caps, . . . grasshoppers performing feats of skill on a high wire after a hearty bowl of Quaker Oats."[54] Because barbed wire was viewed primarily as a way to control animals and because valued livestock risked injury from the wire, it is not surprising that in the industry's very first flier, Glidden's "Farmers! Take Notice," animals offer testimonials on the efficacy of barbed wire fencing (fig. 18). The wire of the two-strand fence is armed with long barbs, plentiful and threatening enough

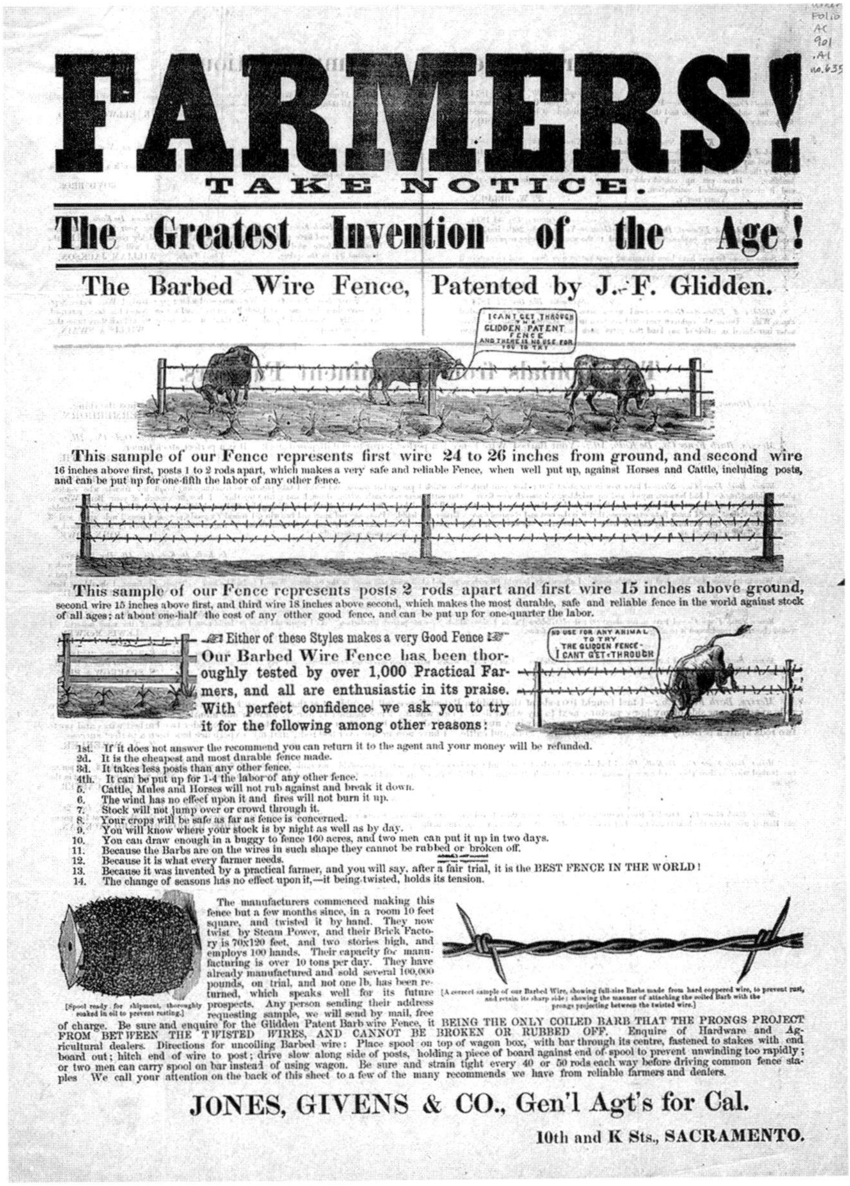

Figure 18. "Farmers! Take Notice." First barbed wire flier, 1875. Courtesy Harold B. Lee Library, Tom L. Perry Collection, Brigham Young University.

Figure 19. Image from a Champion Barbed Wire folding trade card, no date. Courtesy American Heritage Center, the University of Wyoming.

that the bull in the first image explains to the inquisitive but less sagacious cows that if he, with all his male power, cannot get through, they should not even try. In another image, the raging bull repeats that he cannot get through the fence and expands his cow-specific admonition to all animals: "No use for any animal to try the Glidden Fence." The flier suggests to readers that Glidden's barbed wire is more ferocious and more powerful than a fierce bull. As sensibilities about barbed wire changed, however, it became necessary to draw imagined testimony from animals that while the fence controlled their movement, it did so benignly.

One of the more elaborate advertisements employing talking animals is a folding trade card for Crandal Manufacturing. The card presents the ribbonlike "Champion Barbed Wire" fence developed to overcome barbed wire's visibility problems (fig. 19). Behind the fence stands a multitude of barnyard animals. The dog laments his obsolescence, as he is no longer needed to take care of sheep that wisely will not cross the fence. The horse's mea culpa is related to the visibility of Champion barbed wire: "I will be to blame if I get hurt on this barbed wire." And the cow has a plan of action: "I am going down to No. 76 Michigan St., Chicago, and request them to take off some of these barbs so I can get through." There are, then, two strands of meaning in the Champion advertisement: the wire is safely visible (a key issue of the public debate), and the wire is dangerous enough to be effective. When the animals controlled by barbed wire give voice to commonsense logic about its ability to control without harm, who can contradict them?

Haish's early advertising often featured a bellicose fighting cock with which the feisty German American clearly identified. This trade card casts the cock in an aggressive role, but this time it is not a patent cock that suffers. The rooster is the "early bird" that has, as the proverb goes, captured the "worm." The fence is made of boards, and this is the "Eureka" moment when the farmer, who would have noticed the hungry rooster if the fence he was peeing through had been made of barbed wire, realizes he needs another kind of fence (fig. 20). And if a reader thinks the picture is in any way untoward, the trade card turns the judgment back onto the reader: *Honi soi qui mal y pense* (Shame on him who thinks evil). That is not the end of the story, however; the early bird can be controlled and even made useful, as demonstrated on the card's reverse (fig. 21).

Here the renegade rooster must obey the firm commands of a neatly dressed, high-seated, and vigorous young woman. Controlled by barbed wire reins, guided easily by the prim and proper woman, the cock speeds to deliver

Figure 20. "Eureka!" Trade card. Courtesy American Heritage Center, the University of Wyoming.

Figure 21. "Right on Time." Reverse side of the "Eureka!" trade card. Courtesy American Heritage Center, the University of Wyoming.

the "S" Barb wire, stretchers, and staples "right on time," ensuring that no lascivious accidents like the one on the card's other side will occur. In tandem, the two images on this trade card lay out a basic claim: barbed wire may be a little dangerous, but it is certainly dangerous to be without barbed wire. That a barbed wire trade card distributed to farmers, ranchers, and gardeners includes both a ribald joke and a sophisticated French phrase gives a good sense for what Haish calculates are the sensibilities of his potential customers.

The most interesting of Haish's posters features another fighting cock, a magnificent "'S' Barb Rooster" or "Cock of the Walk" kept in check by barbed wire reins in the hand of its gentleman rider (fig. 22). The bird is powerful and threatening with its spurred, twisted-wire legs, assuring the prospective buyer that this fence is dangerous enough to do its job. But not too dangerous; even a pipe-smoking gentleman can control it, especially when aided by barbed wire reins and a barbed wire quirt. The threatening barbed wire, at least in the visual logic of this image, acts as its own restraint. Camels, an elephant, a giraffe, horses, bison, a bull elk, and a couple of jumping frogs populate the poster, exotic and familiar animals securely fenced in by Haish's barbed wire. The exotic elements of the poster add a thrilling context to a scene domesticated by the standing lady and gentleman and the fishing boy. This dangerous-but-benign dialectic is a common theme in Haish's advertisements and in those of his competitors.

Another fascinating set of animals appears in an image Haish used in multiple issues of the *Regulator* and in several posters (fig. 23). The prosperous farm is neatly ordered by barbed wire fences. A train circles around from a factory on a distant hill. Haish's familiar pipe-smoking man wields barbed wire reins and a whip to urge on and control a pair of tigers and a pair of lions. Depending on which version of the image is used as the years pass, "Haish's Trans-continental Through Train" has transported ten thousand or thirty thousand or some other number of spools of barbed wire across the globe. A woman identified in one version of the image as the driver's "Vrow" (*Frau*) stands at a brake wheel on the roof of the second car.[55] The tigers and lions are joined by a bear standing next to a tree wrapped in barbed wire, four pipe-smoking dogs sitting on spools of barbed wire, several squirrels, and an eagle perched on a tight strand of barbed wire above a portrait of Haish. So what does this all mean? Haish's lions and tigers make sense in the context of Lears's argument that "late-nineteenth-century advertising . . . sought both to tame the beast within and tap its vital powers . . . [celebrating] self-control even as

Figure 22. Cock of the Walk. Color poster. Courtesy Regional History Center, Northern Illinois University.

it revealed a powerful fascination with the exotic and the primitive."[56] Haish's exotic beasts are good examples of this double message, both vitally powerful and well controlled by barbed wire reins. The highly civilized pipe-smoking dogs have rifles strapped to their backs, exhibiting their own dual nature. An eagle adds a fierce American note to the scene, and a threatening bear is fired on by the train-riding dog.[57] In short, the Haish flier presents a veritable transcontinental circus under whose tent the viewer witnesses barbed wire as a spectacular synthesis of the exotic and the familiar, of the savage and the civilized.

Like Haish, purveyors of the Glidden fence featured animals of various kinds in their advertisements. In figure 24 the poster's banner title, "Commotion among the Animals," not only bills its product *like* a circus; it bills it *as* a circus. Under the big top, domestic animals easily pass under, over, or through various types of fences as they circle the ring. When they approach the Glidden barbed fence, however, the horses and cows join a lion, an elephant, and a kangaroo that have been thrown into commotion by a fence even they can not

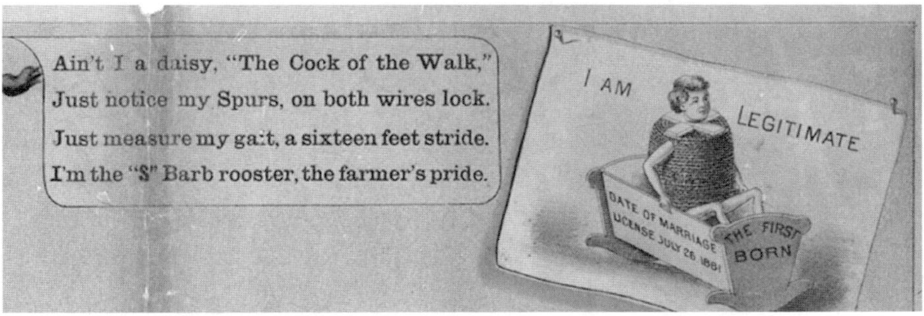

Figure 22 Details. Cock of the Walk. Color poster. Courtesy Regional History Center, Northern Illinois University.

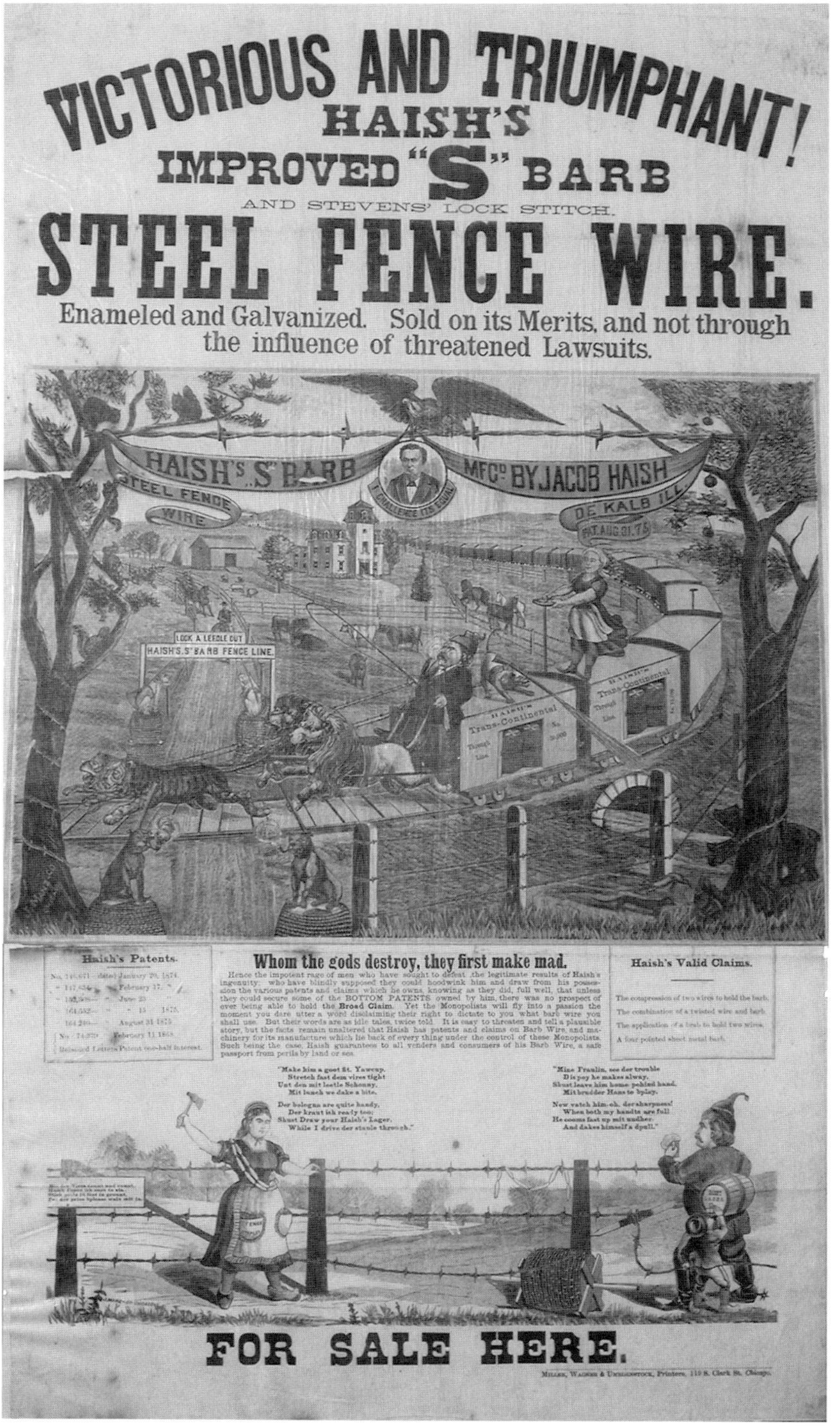

Figure 23. "Victorious and Triumphant!" Color poster. Courtesy the Ellwood House Museum, DeKalb, Illinois.

Figure 24. "Commotion among the Animals." Color poster. Courtesy the Ellwood House Museum, DeKalb, Illinois.

Figures 24 Details. "Commotion among the Animals." Color poster. Courtesy the Ellwood House Museum, DeKalb, Illinois.

jump or break through. However fierce or exotic the animal, Glidden's barbed wire is the master of them all. The revolutionary new fence causes such a commotion that some members of the audience leave the tent in a rush to buy Glidden's barbed wire. The flier is not hesitant to play on national stereotypes, with a Chinese man exclaiming that "barbee wire vellee good, me gettee some too," and a fat German reminiscent of Haish's ubiquitous character saying, "Yaw I thinks I can raise some sourkrout now." The stereotypical portrayals of immigrants reinforce the idea of American ingenuity in solving the fence question. The new American fence, like the telegraph, steam engine, railroad, and plow, helped drive what was popularly seen as "American progress"—and the immigrants are astonished.

The Technological Sublime

In her *Advertising Progress*, Laird writes that "manufacturers . . . boasted of their roles in progress through advertising messages that glorified industry and technology as the means to new abundance and cultural advancement."[58] Haish, Ellwood, and Washburn & Moen all presented images of their factories that bear out Laird's assertion. The multiple massive facades with belching stacks showcase a powerful industry devoted to the new and expanding technology of barbed fencing. We have already noted the factories featured in Washburn & Moen's trade manuals and in the *DeKalb County Manufacturer*. Haish also featured his factory in various publications. One cut presents the busy DeKalb factory and the formidable office as proof of economic potency. Enclosed, protected, and framed by his "S" Barb wire, the factory boasts an erect chimney that raises Haish's name and brand up to where the smoke of industry flies like a banner under Haish's trademark cock. Add the American flag under another commanding cock and depict trains entering and exiting the factory grounds, and you have an image designed to ensure prospective buyers of the quality and secure future of Haish's barbed wire.[59]

A striking Glidden poster features not one, but *three* railroad trains backed by a busy manufacturing plant and flanked by barbed wire fences (fig. 25). The use of the railroad locomotive in advertising is an evocation of the technological sublime, exactly the symbolism that informs this image.[60] While evoking the power of the locomotive (even if "lion-powered"), barbed wire manufacturers present their fence as part and parcel of American progress through technology. Washburn & Moen declared in an 1880 manual that "travellers in all parts of the globe, learn to measure the advancement and progress in civilization of the communities they visit, by the tokens exhibited of care in fencing

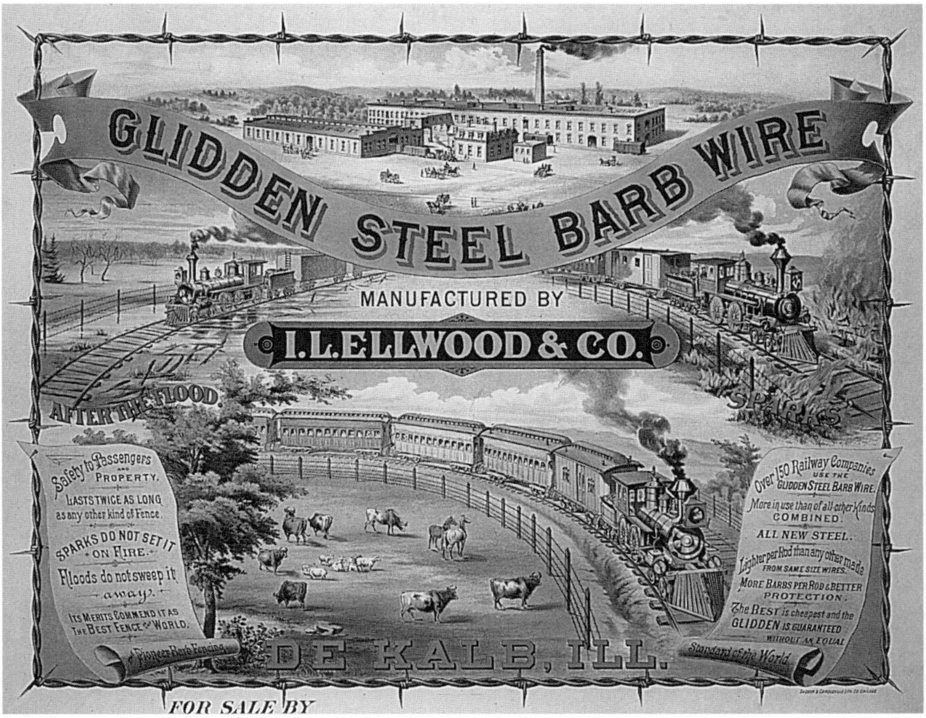

Figure 25. "Glidden Steel Barb Wire." Color poster. Courtesy the Ellwood House Museum, DeKalb, Illinois.

Figure 25 Details. "Glidden Steel Barb Wire." Color poster. Courtesy the Ellwood House Museum, DeKalb, Illinois.

land."[61] Progress, however, especially progress that involves controlling the movement of animals, is part and parcel of progress involving the control of human beings.

Control of Humans

The 1876 Washburn & Moen booklet made the first gesture in that direction. "Growing out of the Barb Fence is the Barb Fence Armor," the text reads, and the accompanying etching illustrates how the coiled barbed wire controls access to the baseball field (see fig. 14). Controlling human beings may have been advantageous for the Base Ball Association, but other kinds of control arose in those early decades that were less benign.[62] In a poetic statement, the 1885 *Glidden Journal* asserts that its fence is a perfect instrument of surveillance: "It watches with argus eyes the inside and outside, up, down and lengthwise; it prevents the 'ins' from being 'outs'; and the 'outs' from being 'ins'; watches at day-break, at noontide, at sunset and all night long."[63] This declaration introduces two related themes: the fencing in of the Same and the fencing out of the Other, and the conversion of the barbs into ever-watching eyes, hundreds of them as in the mythological giant Argus, watching in all directions at all times.

Most notably, argus-eyed barbed wire enabled conquest, control, and exploitation of a vast region occupied by Native Americans. Advertisers were quick to market this aspect of their product. The poster with the wire-legged rooster may be Haish's most troubling image in this context (see fig. 22). Haish's civilized German American astride the double-spurred cock is master of the scene. In the background, under the rooster's barbed stride, stand tiny Native Americans. To the right, disappearing into the background, are several bison and a wagon train. Aided by Haish's barbed wire, the "civilized" Euro-Americans standing at both ends of the fence are destined, the image suggests, to conquer and to civilize. Haish's poster pushes the concept of progress through industry to its utopian (or dystopian) extreme. The Natives and the bison will be overcome by colonial power derived from the invention, manufacture, and deployment of barbed wire. "Civilization" moves forward, the "uncivilized" are forced out, and the driving mechanism is barbed wire. To the left of center, an elephant and a giraffe stand near a seacoast where ships wait to take spools of "S" Barb around the world. The inset top left shows a spry German American atop an Egyptian pyramid enclosed by a barbed wire fence. Haish's barbed wire, a technological miracle and guarantor of civilization, thus controls access to the miracles of Egyptian civilization as well as to the fecund but savage American West.[64]

The general thrust of Haish's fantastic poster would have felt familiar to many Americans. John Gast's 1872 *American Progress*, painted just two years before the first DeKalb patents, likewise depicted the divinely sanctioned displacement of Native Americans by Euro-Americans and their technologies.[65] In Gast's image, the unenlightened Indians are driven into the western darkness by plow-wielding settlers, by an allegorical beauty trailing telegraph wire, and by three trains bringing goods from a busy harbor. Had barbed wire been manufactured in quantity only a couple of years earlier, the painting would certainly have included that technology as well.

The image of the unenlightened savage common to the Haish poster and the Gast painting was widespread. Glidden salesman Henry Sanborn, for instance, planned to make his Texas ranch into a proving ground for barbed wire fences, choosing for the test a "kind of twilight zone with the light of civilization behind it, and the darkness of savagery before."[66] The 1887 *Glidden Barb-Fence Journal* played on Euro-American fears of Natives and the potential for its barbed wire to control them like hogs: "And now the level-headed editor of the Farm Implement News enlarges upon our suggestion of using barbed wire to fence against the Indians. The government should by all means adopt for this purpose the Glidden 'Thick set.' It makes the best 'hog fence' in the world. It might scratch the deviltry out of Geronimo and his gang."[67]

Although the federal conquest of Native Americans was well advanced by the time barbed wire was invented and advertised, the recent defeat of Lieutenant Colonel George Armstrong Custer and his troops at Greasy Grass or Little Big Horn (1876) was still vivid in the minds of potential customers, as were the ongoing battles fought by the Sioux, which ended finally in the winter of 1876–1877, and Geronimo did not surrender for the final time until 1886.[68] Barbed wire was itself a savage tool and thus perfect in the minds of advertisers for controlling "savage" Indians. Not surprisingly, as the real Indian threat subsided, the one dangerous entity came to represent the other. "The Washburn & Moen Mfg. Company," the *Iron Age* reported in 1892, "have just brought out a new pattern of barbed wire. . . . The distinctive mark adopted for it is an illustration of the Indian chief 'Waukegan.'"[69] The wild native, the fictional Chief Waukegan, is introduced, like the fighting cock, as a symbol of strength and ferocity that controlled whatever needed to be controlled. Ironically, the model for Chief Waukegan was a Russian Jew.[70]

The harrowing of corn-ground after planting has grown so much in favor that all who have tried it have adopted the practice. It should be done just before the corn comes through the surface of the ground.

"PEEK-A-BOO."

A DUDELET.

A gorgeous It sailed forth one day,
In brand new suit of colors gay;
And as he walked his wond'rous way
He seemed unto himself to say,
"Oh! I'm too too all but."

Figure 26. "Peek-a-Boo." From the *Glidden Barb-Fence Journal*, 1885.

If barbed wire could control Indians, it could also control recently freed slaves. In the 1885 *Glidden Barb-Fence Journal*, a cartoon played on the set of anxieties that surfaced at the end of the Civil War (fig. 26).[71] Barbed wire has foiled this theft, for no man or beast (and in this racist image the distinction is not great) would dare cross the Glidden fence. The African American man kneeling outside a watermelon patch surrounded by barbed wire fencing is the stereotypical subordinate Sambo figure.[72] The cartoon's "peek-a-boo" caption is an enigma, unless the point is that along with its threatening barbs, the barbed wire is watchful.

Racist references pepper the *Regulator* and *Journal* alike. The 1879 *Regulator* included a cartoon with a slightly more nuanced message. Sambo sits inside a

Figure 27. Sambo and the Bull-Dozer. From the *Barb Fence Regulator*, 1879.

fence surrounding a haystack. Outside stand a jackass and a salesman in a suit carrying a briefcase (fig. 27). Sambo converses with the Bull-Dozer:

> Sambo—It's no use talkin' 'bout yer Mernopoly Barb and Broad Claim; it's dun gone and busted!
> Bull-Dozer—You black rascal! Tell me what you know about it!
> Sambo—Yah! Hay! Massa, when yer solves dis kernunderum fus. How will dat Jackass reach de hay-stack, he can't get under, fru nor ober dat Boss Barb of Haish's?
> Bull-Dozer—I give it up.
> Sambo—So does de oder Jackass.[73]

The Haish cartoon establishes a demeaning hierarchy, using its racial stereotype to skewer the monopolists who were the butt of even Sambo's joke. As noted earlier, calling the salesman a "Bull-Dozer" links him to coercive post–Civil War attempts to keep African Americans from voting. The barbed wire fence, then, protects the often-lampooned Sambo from the racist coercion of the monopolist Bull-Dozer even as it protects the haystack from the other jackass.[74] Lears attributes this kind of advertising to a public uncertain of its own identity in a new commercial era: "All the fears of hypocrisy, all the suspicions that the successful man was merely a clever manipulator of appearances, could be projected outside Anglo-Saxon culture onto the shiftless Sambo, the Indian-giver, the wheedling Semite."[75]

Like his competitors, Haish was not loath to appeal to his readers' racist prejudices. The 1881 *Regulator* leads with Haish's letter to his dealers and consumers: "We need not repeat what you already know, that these selfsame monopolists have persistently published lies that are as black as hell, lies that will damn them as quickly as it will a 'Guinea nigger.'"[76] The same issue of the *Regulator* recounts a story from the *Atlanta Constitution* about Uncle Remus that has him explaining to a black boy with a book and to a policeman acquaintance that education is bad for the boy: "Put a spellin' book in a nigger's hands, en right den en dar' you loozes a plow hand."[77] The story keeps the black boy in his place, as will, by implication, the new fence. Whatever fears circulated after the slaves had been freed, after what was kept in had been let out, barbed wire was promoted as a means to reestablish order.

Advertisements for barbed wire, then, exploited racist sentiments against both Native and African Americans, raising fears and then supplying protection from the threat. In doing so, they were echoing racist advertising for other contemporary products. Robert Jay writes that after the Civil War black people were "viewed as a threat by northern city dwellers as never before, and since most publishers and advertisers were located in northern cities, it is not surprising that throughout the 1880s blacks were treated with varying degrees of derision in the advertising media." According to Jay, post–Civil War representations of African Americans depict them as "simple and good-natured . . . always lazy, often ridiculous," as "pickaninnies" and "watermelon lovers."[78]

That describes the Sambo figure used in barbed wire advertisement perfectly. Two types of Native Americans emerged in late nineteenth-century advertising, media historian Jane Marcellus points out. The "Ignoble Savage" was stereotypically violent, while the "Noble Savage" was the "good" Indian close to nature. The Native Americans used in advertisements for barbed wire are not romanticized symbols of wisdom like those used in advertisements for patent medicines; instead, they symbolize the noble savagery of the barbed fence. Mirroring our own sense for the use of post-conquest Indians in barbed wire advertising, Marcellus reasons that "the appropriation of the past reflects a need . . . for the dominant culture to be reconciled with, and perhaps healed by, that which it had subjugated."[79] In regard to the stereotypes used in nineteenth-century advertising, psychologist Arthur G. Miller argues that "the major function of attaching labels to different racial and ethnic groups is to impose order on a chaotic social environment."[80] Barbed wire advertisers doubled down on this function, as their product itself was a potent means for imposing order.

Advertisers were quick to pillory other groups as well, people against whom they could assume the buyers of their wire were prejudiced. "A Dudelet," for instance, a beautifully written if sinister poem printed directly under the "Peek-a-Boo" cartoon in the *Barb-Fence Journal* (see fig. 26), plays on a prejudice against "dudes," or aesthetes like Oscar Wilde, who toured America in 1882.[81]

> A Dudelet
> A gorgeous It sailed forth one day,
> In brand new suit of colors gay;
> And as he walked his wond'rous way
> He seemed unto himself to say,
> "Oh! I'm too too all but."
> . . . .
> Passing along he met a That,
> With figure fine—though most too fat—
> Button-hole bouquet and soft crush hat,
> White kid gloves and flash cravat,
> A most too aesthetic mortal—quite.
> . . . .
> They bowed and scraped, first toe, then heel,
> 'Till It stepped on an orange peel.
> Oh! Spare your smiles, for pains they feel,
> They both got pricked on barbs of steel.
> "The hor-wid bar-b'd wi-ah."
> . . . .
> Beware the peel—and Glidden wire—
> They're death on dudes.[82]

Harvard students are mocked in the 1885 *Glidden Barb-Fence Journal* with a recipe for their manufacture: "Take a lump of egotism, add a slice of skepticism / Mix them well together with a 'culchared' Boston drawl / add a little Darwinism, just a smack of positivism / and flavor with the essence of unmitigated gall."[83] Glidden's 1881 *Journal* appeals to anti-intellectual bias to attack its antimonopoly opponents:

> The men who are now howling "monopoly" are the same chaps who a few years ago were crying down the use of Barb Wire as being barbarous, unsafe, etc. These men are constantly being crowded to the wall by the march of events. The fact is, they are anxious to air

> their philosophy in the public prints, and pose before the people as the "farmers champions." Their theories are fine, their logic pointed, and their periods well rounded, showing themselves masters of elocution and of the pen. Perchance they may be "professors" of something or other, but evidently they know as little of the great industries of the country in their practical application, as "cows do about shoemaking."[84]

These mocking depictions of effeminate, educated city slickers were aimed at the journal's largely rural and male readership. So were the frequent jokes about women.

The *Barb-Fence Journal* regaled its male readers with quips like this one: "'No abridgement of the liberties of the press,' as the editor said when he squeezed the hired girl."[85] A long story about a man who followed a newspaper's advice that he should "treat your cows as you would a lady" but was kicked and gored despite wearing his suit and politely addressing one of them, ends with him canceling his subscription to the paper.[86] "A Zulu belle," the *Journal* reports, "may be said to be like the prophets, because she has not much on–er in her own country."[87] The same volume points out that "old maids are embers from which the sparks have fled" and recounts a tobacco-related story with the title and punch line "Rather Chew Than Marry."[88] The subsequent volume reports a conversation between two men: "How's your wife, John? Who? Your wife. Oh! She's well. Thought she was sick and goin' to die. Yes, we all thought so but she disapinted us."[89] Haish's *Regulator* also titillates its readers with stories like the one that spells out a man's disgust with a woman practicing calisthenics and his threat to beat her with a broom unless she gets to the washing tub quickly: "Calisthenics, ha! I'll have no more of this jig-jaggin' around like a hin and a stove-lid. The foorst thing ye know it's joinin' the bally she'll be, an' be spendin' all yer money for clothes an wearin' none of 'em. Calisthenics, O ho!"[90]

The white, male, robustly heterosexual farmers and ranchers who might buy the "S" Barb or the Glidden Coil found their prejudices against Indians, blacks, aesthetes, and women reinforced in the *Regulator* and the *Journal*. They could act on those prejudices, Haish, Ellwood, and Washburn & Moen proposed, by buying the wire that was "death on" anyone not their own kind (hardy, white, and masculine members of the benevolent and vicious order of the erectors of fences). Barbed wire's ability to control errant (by whatever standard) human behavior resulted, then, in standard caricatures employed by advertisers: savage Indian, black thief, gorgeous aesthete, "uppity" women.

Whatever the threat, barbed wire was the solution. And where solutions were so absolute, a turn to religion was inevitable.

Religion

As noted earlier, Haish's 1877 *Regulator* featured a vivid image of a horned "Father of Lies" with the "Broad Claim" rising from his back pocket and Glidden's much-touted "bevel cut" stamped on his cloven foot (see fig. 15). Frustrated after the 1880 court decision in favor of Glidden's claim to a broad, previous patent and against his own patent, Haish began the 1881 issue of the *Regulator* with a paean to God-fearing dealers and consumers "whose convictions of right and wrong are based upon . . . an All-wise and truth loving God."[91] Besides such direct attacks on competitors as Godless devils, Haish variously employed religion to sell his wire, proclaiming in volume 6 of the *Regulator*, for instance, that his invention was the "Pioneer, Light-weight Wire of the universe, also for the promulgation of the true creed of Barb Wire Gospel, stamped with the sanction and authority of the wisest and best throughout the civilized world."[92] The same issue of the *Regulator* concludes with a sermon promising salvation from the cares of the world: "The weary world has need of rest. It seethes and boils like the witch's cauldron. . . . What means this unrest; this fever heat? . . . Who is the physician able to . . . effect a certain cure . . . ? The cure is simple; it is within the reach of all . . . : Buy the celebrated Enamel 'S' Barb Steel Fence Wire, manufactured only by Jacob Haish."[93] Similarly, the 1887 issue of the *Regulator* includes a poem promising revealed truth and ending with a rhetorical flourish:

> For the truth is here concealed,
> But unto you it is revealed,
> That "S" Barb for your field
> Is your protection, guard and shield.[94]

Haish repeatedly takes a sacred motif that might be seen as off-limits to commerce and, tongue in cheek, inserts his fence in the place of God. In her essay "Advertising—Sacred and Profane," Marghanita Laski suggests that this is a treacherous choice. Potential consumers of the product might react badly to barbed wire invasion of their sacred rhetorical space. Roland Marchand cites and then continues Laski's argument: "To shout 'Be patriotic; buy my product,' or 'Experience the sublime rapture of owning a product illuminated by God's holy light; send your money today,' would embarrass any respectable

Figure 28. "Forbidden Fruit." Trade card. Courtesy American Heritage Center, the University of Wyoming.

advertiser."[95] It obviously did not embarrass Haish, who calculated that his folksy humor would be appreciated by his barbed wire–stretching audience.

The salvific nature of barbed wire is reiterated on one of Haish's trade cards: "Forbidden Fruit. Can't get through Haish's 'S' Barb Fence."[96] Standing in a proverbial Garden of Eden with animals grazing peaceably around her, a young woman picks fruit and gathers what she picks in her upraised dress. In the lane on the other side of the barbed wire fence stand two men who have tried to reach the forbidden fruit (the girl? the apples?). A third man has turned to run (fig. 28). The awkward stances, the possibly injured hands, and the man's flight indicate that the attempt to transgress the commandment has been duly punished. Adam and Eve may have eaten the "forbidden fruit," but that was before the deployment of barbed wire. In stark contrast to the sweet girl picking apples are the fence barbs, so exaggerated and dominating that the bucolic landscape takes on a threatening air. Who would want to live in the midst of so much dangerous wire? But then again, who would not want to live in a peaceable kingdom safe from transgressors? Yes, barbed wire is dangerous, and because it is dangerous, it protects the innocent.

The *Glidden Barb-Fence Journal* likewise exploited religious imagery and language to reach its audience, as in the following examples. "Did the prophet Isaiah ever eat at a railroad station? It certainly looks so, for how could he have described it so literally if he had not: 'And he shall snatch on the right hand and be hungry; and he shall eat on the left hand, and shall not be satisfied.'"[97]

> A clergyman asked his Sunday school: "With what remarkable weapon did Samson at one time slay a number of Philistines?" For a while there was no answer, and the clergyman, to assist the children a little, commenced tapping his jaw with the tip of his finger, at the same time saying, "What's this? What's this?" Quick as thought, a little fellow immediately replied; "The jaw-bone of an ass, sir."[98]

As with other marketing strategies, religious references are meant to draw in potential customers, in these cases customers with religious backgrounds and without overweening piety. Advertisements also drew on a set of images associated with the home, sure to tug at the heartstrings of potential customers and reflecting the security and orderliness created by established class and gender roles.

Ordering the Landscape

The letter introducing the 1887 *Regulator* ends with a tongue-in-cheek "regulatory" forecast: "We have also undertaken to Regulate the Eclipses in 1887 as well as some other matters, and the following [schedule of eclipses] is the true and correct style of eclipse which will be right on time as predicted. Yours Astronomically, J. Haish & Co."[99] If Haish's barbed wire can promise order, true and correct order in the universe, then surely a smaller matter such as providing domestic security will be an easy task.

The cover of volume 11 of the *Regulator* presents a vignette of rural America quite striking in its detail (fig. 29). We see a farm set back from a tidy roadway on which a figure stands in front of a closed property gate. The residents have secured the house's immediate surroundings from encroachment with barbed fencing. The fence cuts across the roadway, neatly dividing the land into protected interior space and open exterior. Beyond the gate, the road curves past a multitude of trees, and a large farmhouse and barn stand in the background. This is a prosperous and well-ordered farm. There are no animals visible in the house's immediate environs, indicating a clear division between human and animal space, a sense reinforced by the bull (or is it a cow?) standing outside the barbed fence. The fence barbs are visible, although not exaggerated in any way that would feel threatening. The figure in the foreground, returning from fishing, carries his pole and wears a small pack labeled "S Barb" in

Figure 29. Cover of *Barb Fence Regulator*, 1887. Courtesy American Heritage Center, the University of Wyoming.

reference to Haish's most successful patent. To the figure's right is a full roll of barbed wire ready to deploy. Watching over the whole scene is the mustached, pipe-smoking gentleman whose eccentric soft cap we have seen so often before. Details of the textual design are interesting as well. The font for "REGULATOR" is generally rounded, while the font for "The BARB FENCE" is sharply pointed. The engraver (Jacob Manz & Co.) thus presents a combination of dangerous and benign signifiers. Haish's "S" Barb is just dangerous enough to produce comfortable security for families and homes.[100]

Like the *Regulator* cover, the front covers of the *Glidden Barb-Fence Journal* depict a peaceful rural scene carefully ordered by barbed wire fences (fig. 30).[101] Neighboring farms are separated by barbed wire; each of the homes is encompassed by barbed wire; the railroad right-of-way is fenced on either side to protect travelers and animals alike. The impression is of orderliness, tranquillity, security, and prosperity. What should be in is in, and what should be out is out. Animals stand or lie peacefully behind barbed wire fences, apparently familiar with the consequences of confronting the barbed wire. Protected by barbs, the humans nestled inside the fence are snug and secure, their peace and prosperity assured.

With these images in mind, it is interesting to note that at least as early as the first half of the nineteenth century, farm fencing was thought to reflect the character of the individual farmer. In his study of early nineteenth-century agricultural treatises promoting the adoption of live hedges as enclosures, Paul Bourcier found that fencing proponents couched their advocacy in terms of private and public virtue.[102] Well-maintained hedges reflected an individual farmer's industry, vigilance, patience, skill, and wealth, all characteristics required to grow and maintain live-hedge enclosures. Later in the century, horticultural organizations like the Iowa State Horticultural Society decried the encroaching barbed wire (and perhaps the erosion of class) with reference to hedges:

> I desire to call your attention to a return to the growing of Osage orange for fence purposes, with a view of abolishing barb wire as much as possible. Barb wire fencing is in no way cheaper than Osage hedge plants, and the value when made is in my judgment two hundred per cent less than a hedge, both being made in the best manner. When once erected, wire is almost as expensive to keep in repair. . . . The damage to both man and beast annually will keep trimmed and in the best of repair a hedge fence from one year to another. And what is more beautiful? An ornament in summer, a wind-break in winter, and an everlasting fence, care for it or not. Save the cost of plants to the builder (twelve dollars to the mile), the farmer works in his time and saves remaining cost at home. Much more can be said in condemnation of barb wire; then what will we conclude?[103]

Bourcier notes that visitors to early nineteenth-century rural America saw how fencing "tamed" the agricultural landscape through orderly divisions within and between farms, creating "beauty" through "regularity." The concept of fencing itself reaffirmed ideas of protection and security as well as the right of private property by delineating boundaries that promoted "good fellowship among neighbors and encouraged peace and civility among landholders."[104] The ideas of security, order, wealth, and virtue associated with living hedges gained new life in the second half of the nineteenth century via the advertisements for an "artificial hedge" (barbed wire), as in this assertion in the *Glidden Barb-Fence Journal*: "The character of a man may be known by a glance at his surroundings. If the farmer's buildings be snug . . . his stock sleek, well fed and kept in their proper places by good fences, you may be morally certain he is a prosperous, well-to-do and influential man in his community. You may also be certain that he used the Glidden wire."[105]

In order to corner the fencing market, barbed wire manufacturers derided the use of Osage orange and other live hedges. The *Glidden Barb-Fence Journal* was direct in its attack on the very live hedges that had been viewed earlier as evidence of character: "Barbed Wire has come to stay. Hedges must go."[106] And "a farmer who can sit down and wait for a hedge to grow must have the patience of Job, and more leisure than most farmers have."[107]As barbed wire manufacturers played up the problems associated with living hedges, they also, ironically, likened their product's qualities to those previously attributed to hedges:

> A high barbed wire fence composed of many strands . . . is highly ornamental for gardens, parks, orchards, graperies, flowerbeds, private grounds or lawns, and impassable by man or beast; and is the only fence or barrier that obliges every man or animal to enter the premises through the gate. Vines cling naturally to such a fence, and present a beautiful living hedge.[108]

Such accounts work on two levels. First, they break down class-related agricultural methods that relied on the use of wood, stone, or Osage orange hedges—all requiring substantial time to achieve and continual resources to maintain—and they do so by offering a cheaper technology not requiring the resources of previous fences. With barbed wire, the masses could now fence their property as readily as well-to-do gentleman farmers. Second, they employ images of large, prosperous farms surrounded by barbed wire. All

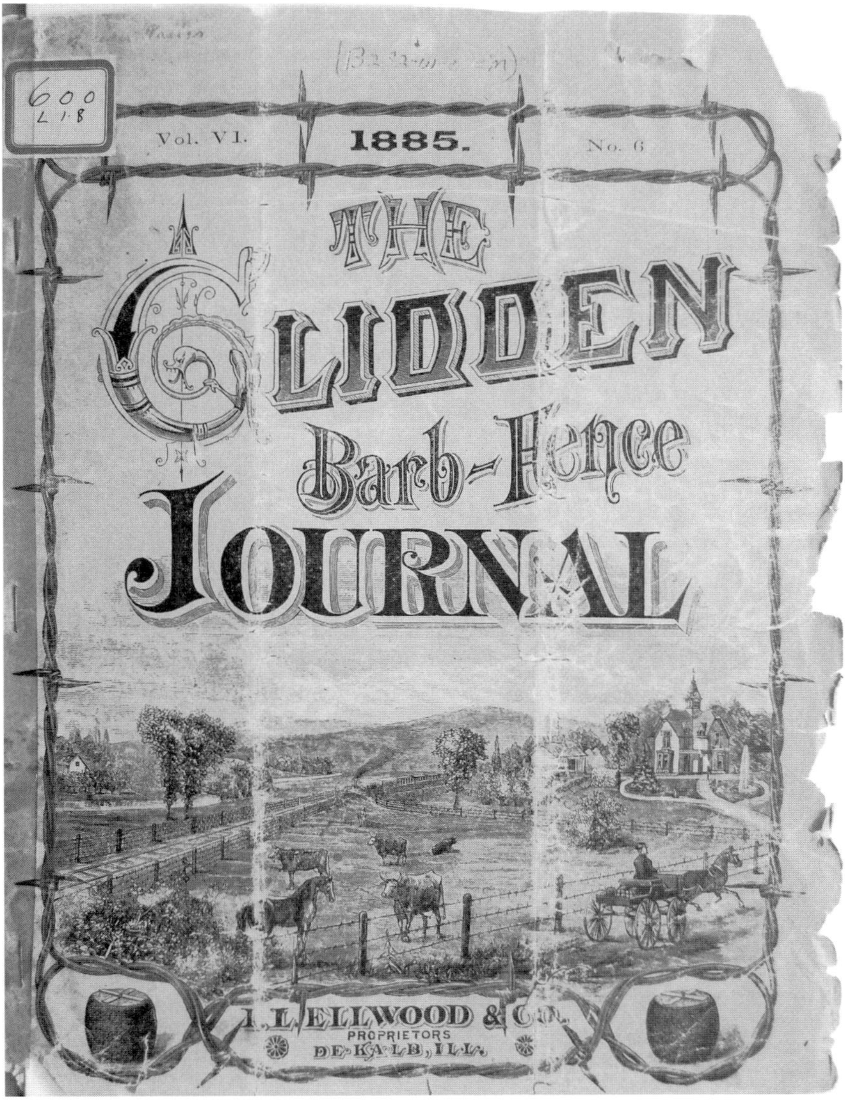

Figure 30. Cover of *Glidden Barb-Fence Journal*, 1885. Courtesy American Heritage Center, the University of Wyoming.

farmers intelligent enough to fence with barbed wire could thus transform their property into the beautified spreads that testified to the high character an earlier generation had attributed to hedges.

Images that tout the order provided by barbed wire are a stock in the trade, like the two images juxtaposed in the *De Kalb County Manufacturer* discussed earlier (see figs. 6 and 7).[109] Chaos reigns in the first drawing. The accommodations in the clearing are primitive, the animals are not controlled by the inadequate split rail fences, and the amount of work that must be done is insurmountable. It is a landscape in need of domestication. All that is necessary to make the leap is, of course, the introduction of barbed wire. The new barbed system depicted in the second drawing establishes domestic security and prosperity as it exerts its ordering influence.[110]

White picket fences, like the live hedge, were idyllic enclosures for nineteenth-century rural and urban homes. The ornamental fences both delineated property lines and established a domestic atmosphere. The 1867 patent from Alphonso Dabb, while not incorporating barbed wire per se, applied "ornamental" cast-iron pickets to preexisting fences, giving them "the advantage of great strength, . . . security . . . and ornamental appearance"—exactly the threatening, protective, ornamental features barbed wire afforded.[111] Royal Eaton's patent for a "Barbed Picket Fence" that was "strong" and "ornamental as well" is a traditional domestic picket fence spiked with barbs. Four years later, James Locher's 1888 patent for a picket-and-barbed fence also included a diagram of a traditional picket fence armed with metal barbs. Perhaps the best example of melding domesticity and security appeared in 1890 with Henry Volck's patented "Picket Fence." Volck's "new and useful" improvement was simply a fence built with wooden pickets embedded with barbs that were strung along twisted double wire.[112] Unlike traditional picket fences one could lean over to chat with a neighbor, pick a flower, or retrieve a ball, this picket fence was to be avoided; outward-facing barbs kept the world at bay, the domicile secure.[113] What is striking about these barbed picket fence patents were the inherent contradictions they represented. Wooden picket fences reinforced concepts of civility and domesticity. The addition of metal pickets or steel barbs of various design promise to enforce domestic serenity and security, which are to some extent gender specific, especially when coupled with beauty.

In her survey of early nineteenth-century women's magazines, Barbara Welter includes domesticity among the virtues of a contemporary woman:

> The attributes of True Womanhood, by which a woman judged herself and was judged by her husband, her neighbors and society could be divided into four cardinal virtues—piety, purity, submissiveness and domesticity. Put them all together and they spelled mother, daughter, sister, wife-woman. Without them, no matter whether there was fame,

achievement or wealth, all was ashes. With them she was promised happiness and power.[114]

A woman's place was in a home safely secured from the outside world. And what could possibly protect women and reinforce their domestic roles better than barbed wire?

A particularly savory version of German American domesticity was promised by an image that recurs in Haish's *Regulator* and his fliers alike (see fig. 23). Stringing the dangerous wire is easy enough to allow conversation about a good German lunch and joking about Jimmy who pulls at the beer keg on Vati's back while the supremely domestic couple goes about its pleasant work:

Der bologna are quite handy,
Der kraut ish ready too;
Shust Draw your Haish's Lager,
While I drive der staple through.[115]

This version of the verse is taken from the 1875 Haish advertising poster "Victorious and Triumphant! Haish's Improved 'S' Barb." Haish published other versions of this cut with the word "soda" replacing "lager," a nod, perhaps, to the contemporary influence of the Women's Christian Temperance Union. As we have seen, Haish often presents these genial Germans as embodiments of domestic prosperity secured by the admittedly dangerous wire. Here they stretch barbed wire quite nonchalantly. What is there to be afraid of? This substantial man, this sturdy woman, and their ingenious fence will regulate your life, providing prosperity and domestic security in a context of beery humor.[116]

Consolidated Steel & Wire Company's 1895 "Pocket Record" features a bucolic scene with a farmhouse and an apple orchard surrounded by barbed wire. In that serene context, a young boy flees the scene of a crime. The story is clear: The would-be thief trespassed, gathered apples in his hat, and attempted to flee from the resident dog. Unfortunately, this misguided youth misgauged the ease with which he could cross through or over the barbed wire. He has hooked his pants and jacket on the fence barbs and spilled his cargo. The slogan, "'Baker Perfect,' Will Catch 'Em Every Time . . . Who will care for Mother Now?," speaks to the dire consequences of jail, injury, and separation from a needy mother. The company describes its Baker Perfect barbed wire as "safe to use," except, of course, for thieves (fig. 31). The Pacific

Figure 31. Consolidated Steel & Wire Company "Pocket Record," 1895. Courtesy American Heritage Center, the University of Wyoming.

Figure 32. Cover of Pacific Steel & Wire Company, *Fencing For All Purposes*, 1897. Courtesy California History Room, California State Library, Sacramento, California.

Steel & Wire Company's 1897 wire fencing circular also advertises domestic applications. The front cover includes typical farm animals to be fenced in, in this case a sow lying contently as she suckles her litter within the confines of a wire fence (fig. 32). To the right of the sow and piglets, a parallel human scene is depicted. Barely visible through tall trees stands a stately Victorian home. In a meadow, a woman reclines in a chair with book in hand as two of her offspring play ball at her feet. Like the sow and her litter, this mother, her children, and their domicile are securely fenced in. The implicit equation of women and pigs has a corollary in the maudlin stories about mothers and wives that appear repeatedly in both barbed wire almanacs. A story titled "They Never Strike," for instance, appeared in the 1887 *Regulator*:

> There is a class of people in this country who get up at 5 o'clock in the morning and who never get back into bed until 10 or 11 o'clock at night; who work without ceasing the whole of that time, and receive no other emolument than food and the plainest clothing; they understand something of every branch of economy and labor, from finance to cooking, though harassed by a hundred responsibilities, though driven and worried, though reproached and looked down upon, they never revolt, and they cannot organize for their own protection. Not even sickness releases them from their post. No sacrifice is deemed too great for them to make, and no incompetency in any branch of their work is excused. No essays or books or poems are written in tribute to their steadfastness. They die in the harness and are supplanted as quickly as may be. These are the housekeeping wives of the laboring men.[117]

While these kinds of stories have no reference to barbed wire other than the fact that they appear in barbed wire almanacs, they put women in their place the way barbed fences enclose animals. Here women are explicitly compared to draft animals who die in the harness. Here and elsewhere the women are so glorified that good farmers reading the almanacs might well be prompted to buy barbed wire to protect their women as they do their prize pigs. The references to women reinforce the central tenets of True Womanhood that the producers, distributors, purchasers, and users of barbed wire accept as the natural order of things.

As barbed wire advertisements use images and stories promising to control Indians, freedmen, effeminate Others, and women; as they assert their product as another example of the technological sublime related to American progress; as they exploit religious belief; as they equate character and virtue to barbed wire fencing; and as they promise prosperous domesticity, these manufacturers sell their products by targeting an array of prejudices, moral positions, and worldviews. Each of the advertisements discussed here shares the precarious dialectical message required of all barbed wire advertising: This wire, this sharply barbed wire, this very effective wire, may be dangerous, yet it is benign.[118] If exponentially rising barbed wire sales in the late nineteenth and early twentieth centuries are any indication, the message was well received.

PART TWO

The Barbed Wire Motif in Literature

Metaphors offer to change the world by changing one's sense of it.
The best metaphors are revolutionary.
—Denis Donoghue, *Metaphor*

I was convinced that cultural theory and literary criticism needed a comparably new idiom, beginning with the effort to think with or through the physical object world, the effort to establish a genuine sense of the things that comprise the stage on which human action, including the action of thought, unfolds.
—Bill Brown, *A Sense of Things*

Barbed wire is a manufactured object. It is also an object whose initial meanings were constructed through public debate and extensive advertising. As noted previously, the value and consumption of a commodity like barbed wire, according to Arjun Appadurai, is "subject to social control and political redefinition." "The politics of value," Appadurai argues, is a "politics of knowledge."[1] Nowhere is that constructed knowledge more evident than in literature. The barbed wire that comes to inhabit works of literature is a "thing" in the sense Bill Brown addresses as he explores "why and how we use objects to make meaning, to make or re-make ourselves, to organize our anxieties and affections, to sublimate our fears and shape our fantasies."[2] These are our questions as well. How, we ask most broadly, has barbed wire transformed what it means to be human? Barbed wire is meant to control. Advertisers worked to make it *mean* control as well. Poets and novelists found the metaphor irresistible.

Barbed wire entered literature in the poetry used to tout the product. Mark Twain's 1889 *A Connecticut Yankee in King Arthur's Court*, published just fifteen years after the new fence began to be manufactured and sold, explored the meaning of barbed wire from a vantage point outside the realm of advertising.

Twain's satirical novel is aimed squarely at his product-obsessed and advertising-assailed American contemporaries. Hank Morgan, the Connecticut Yankee, points out that "many a small thing has been made large by the right kind of advertising."[3] He dispatches his knights as aggressive salesmen offering products that epitomize nineteenth-century "civilization":

> Their penchant for wandering, and their experience in it, made them altogether the most effective spreaders of civilization we had. They went clothed in steel and equipped with sword and lance and battle-axe, and if they couldn't persuade a person to try a sewing-machine on the installment plan, or a melodeon, or a barbed-wire fence, or a prohibition journal, or any of the other thousand and one things they canvassed for, they removed him and passed on.[4]

Late in the novel, when Hank's civilizing endeavors and his life are threatened by an army of angry knights, he fortifies his position with electrified wire—originally meant by Twain to be electrified barbed wire.[5] The scene is horrific, prophetic of World War I trench warfare dominated by barbed wire entanglements (which slowed down advances) and the machine gun (which could kill more efficiently than rifles).[6]

> Yes, it was a man—a dim great figure in armor, standing erect, with both hands on the upper wire—and of course there was a smell of burning flesh. . . . We counted fifteen of those pathetic statues—dead knights standing with their hands on the upper wire. . . . The thirteen gatlings began to vomit death into the fated ten thousand.[7]

Twain's intuitions about barbed wire as a tool of war became reality in the Spanish-American War, in the concentration camps of the Boer War, in the trenches of World War I, in the extermination camps of the Holocaust, in US internment camps for Japanese Americans, and in Soviet gulags. In language that echoes Twain's, Erich Maria Remarque depicted the military horrors of barbed wire in his novel *All Quiet on the Western Front* (1929): "I see one of them, his face upturned, fall into a wire cradle. His body collapses, his hands remain suspended as though he were praying. Then his body drops clean away and only his hands with the stumps of his arms, shot off, now hang in the wire."[8] DeKalb folklore suggests that Jacob Haish, appalled by the human cost of the war, refused to sell his barbed wire for use on the battlefields of World War I and closed down all production in 1916.[9]

In a mundane sense, barbed wire was such a novelty that authors more than a century later can still evoke the late nineteenth century by reference to the fence. Early in Luis Alberto Urrea's 2005 novel *The Hummingbird's Daughter*, for instance, Tomás Urrea sorts through goods delivered to his Mexican rancho, several of them inventions that place the action in the 1880s: "A Singer sewing machine. . . . One slightly rotten burlap bag of the new Burbank potatoes. . . . A sample board of ten of the new barbed wires from Chicago."[10]

Twain, Remarque, Urrea, and many others have used barbed wire in these sorts of historical contexts. Barbed wire is also a ready metaphor. The movement from thing to metaphor is made explicit in video artist Bruce Nauman's *Setting a Good Corner (Allegory & Metaphor)*.[11] In 1999, Nauman filmed himself putting up corner posts for a barbed wire fence at his New Mexico ranch. Not just any corner, but a good corner without which tight-stretched barbed wire will sag. Nauman emphasized the connection between the real fence and the structure of a work of art: "In this case, we're building a corner to stretch a fence and hang a gate. It had a real purpose on the ranch here; I needed to do this. But at the same time, it made a beautiful structure, and it also was one of these things: 'Well, here's the job and . . . '"[12] It is this "and" that will be our focus in the chapters of part 2.

Two very different poems with barbed wire as a central motif set the stage. Both reference the history of the fence and highlight the control barbed wire was invented to exert. The major difference between the poems, other than quality, is that the first is from the perspective of those who are striving to control while the second perspective is of a person threatened by control.

Elliot Lincoln claimed a Montana homestead in the first decade of the twentieth century and over the next years published folksy poetry about frontier life in magazines and in two books. His poem "A Song of the Wire Fence" depicts the fencing of wild lands as American progress and as the heroic subjugation of nature. With its lofty tone and wire-fence persona, it stands out from the colloquial and often humorous stories of the other poems in the book *Rhymes of a Homesteader*.

> Millions of miles of shining metal threads
> Cutting the plain in geometric lines,

Climbing aloft among the mountain pines,
I show the way wherever Progress treads.

I bound the cultivated fields of man,
Divide his cattle from the masterless,
I form a barrier to the wilderness;
I end that which has been since time began.

My barbed and twisted strands have marked the change
That comes when Nature pays the debt she owes.
I whisper to each heedless wind that blows
The last low dirges of the open range.[13]

The wire's enthusiastic self-praise is reminiscent of paeans to the technological sublime (the steam engine, the railroad locomotive) in the previous century. The poem mates a rather simple technology with the triumphant rhetoric of conquest and cultivation. The geometrical overmastering, bounding, cutting, dividing, barricading, and marking of nature is depicted as progress with a capital "P." The shining barbed wire has structured a wilderness and forced Nature to pay a debt she owes. Here, as in the traditional (Turnerian) narrative of the Great Plains, Native Americans are noticeably absent, the impact of technology on their cultures unconsidered.

Amy Clampitt's "The Woodlot" reaches back into her Iowa prairie childhood. The poem begins with lines about configuring the prairie into grids that echo Elliot Lincoln's "geometric lines":

Clumped murmuring above a sump of loam—
grass-rich, wood-poor—that first the plow,
then the inventor (his name plowed under
somewhere in the Patent Office) of barbed wire,
taught, if not fine manners, how at least to follow
the surveyor's rule, the woodlot nodes of willow,
evergreen or silver maple gave the prairie grid
what little personality it had.

Where Lincoln's barbed wire forced itself on a nature that had no choice but compliance, Clampitt introduces a force of nature of a less compliant sort, the shape-shifting air that is "oblivious of section lines," indicate ownership. The fence can control some entities, but is powerless against the storm:

. . . Barbed wire
kept in the cattle but would not abrade
the hide or draw the blood
of gales hurled gnashing like seawater over fences'
laddered apertures. . . .

Cattle are one thing, but when the air takes on the shape of a tornado, the fence is of no use, "luck and a cellarhole were all / a prairie dweller had to count on."

The shifting forces of nature and the accoutrements of fixed human settlement are antagonists. Clampitt's poem asks whether the inventor of barbed wire predicted the extent of influence his fence would exert:

. . . Did he
ever, even once, envision
the spread of what he'd done
across a continent: whale song's
taut dulcimer still thrumming as it strung together
orchard, barnyard, bullpen, feedlot,
windbreak . . . ?

The ingenious inventors of barbed wire did in fact envision how the continent would be changed with the stringing of their wire. But this poem is less about the effect of barbed wire on the continent than about how it shapes the life of the poet. The real question of the poem is whether the inventor ever envisioned the ability of barbed wire to control the possibilities of a woman's life. When the poem's prairie dweller escapes the confines of the barbed wire ("wire to be clambered over, / crawled through or slid under, shepherded— / the heifers staring"), she finds, in the woodlot, a natural opening, a place of safety, refuge not from a tornado but from the strict linear ordering of the fence:

. . . Deep in it, under
appletrees like figures in a ritual, violets
are thick, a blue cellarhole
of pure astonishment.[14]

This blue cellarhole in the woodlot is "the earliest memory," a memory mercifully separated from the girl's experience with the constructions and inventions and tools that we use to force our will on each other and on the world we inhabit.

This final turn is a critical moment for understanding the poem and its barbed wire metaphor. The historical right-angled sectioning of the prairie by surveyors and the fences of the prairie that enable farming enclose the person as well as the heifers. In an essay called "Providence," Clampitt describes the upheaval caused by her earliest memory:

> It is the twenty-sixth of April, 1923, I am not yet three years old. . . . My father's sister Edith . . . is leading me past barns and through feedlots to the outermost grove. What holds these details in place is the sight, out under those trees, of a bed of violets whose hue I cannot reach except by way of a later metaphor: the contained intensity of a body of water. It is as though I became in that instant aware of edges, shores, boundaries, limitations. The shell had cracked: an exodus, an expulsion, was under way."[15]

These boundaries and limitations find material form in the barbed wire of the poem. The fixed edge, the barbed enclosure, and the brave but ultimately tenuous attempt to order and control the prairie and, metaphorically, the poet's life, fall away at the sight of the violets whose hue reminded the poet of water. The blue violets below and the watery, shape-shifting air above crack the shell of the girl's prairie identity. She can be another sort of person. The strict requirements of prairie farming can be exchanged for life in a city, for travel, for anything beyond the barbed, twisted-wire fences that make a prairie life possible and that, while doing so, impose their structures on the people who use them.

In rhetorical terms, Bruce Nauman, Elliot Lincoln, and Amy Clampitt all use barbed wire as the vehicle of metaphor. That vehicle—the thing itself and the control it exerts—has the capacity for a wide range of meaning. Alan Krell's "cultural history of barbed wire," *The Devil's Rope*, explores many of the meanings barbed wire has acquired over the decades following its invention."[16] His mostly visual examples are ordered thematically, as his chapter titles suggest: "Tortured Bodies / Touching Sites," "Making Familiar," "Entangled Intimacies," and "Grasping the Nettle." In part 2, we expand Krell's groundbreaking cultural history with a focus on works of American literature and film. In the texts we examine, barbed wire is more than a passing reference, more than a historical locator; it is a critical contributor to the meaning of the work.

Addressing problems in the analysis of painting and sculpture, Erwin Panofsky lays out a process related to our work: "*Iconographical analysis*, dealing with *images, stories* and *allegories* . . . presupposes a familiarity with specific *themes* or *concepts* as transmitted through literary sources."[17] The question of what barbed wire means in any specific work can be answered only with the historical context as background and in the complex contexts of the works themselves. Literary motifs are "the elements from which traditional narrative literature is constructed," according to Theodore Ziolkowski, and "themes, motifs, and images constitute an important link between . . . literary work and the social, cultural, and historical contexts."[18] Further—and this is certainly the case in our own explorations—"the sensible application of thematic analysis can expose aspects of the work itself that would otherwise remain inaccessible."[19]

In the following chapters we examine works of American literature and film that exploit the barbed wire motif. Chapter 4 examines genre Westerns, including works by Mollie Moore Davis and Owen Wister. Chapter 5 surveys fiction of writers of the New West such as Edward Abbey, James Galvin, and Annie Proulx. Chapter 6 explores flagellant varieties of religious life in prose fiction by John Steinbeck, Flannery O'Connor, Brian Evenson, and Jeff Mann. And Chapter 7 raises questions of Native American and hybrid identity in works by D'Arcy McNickle, Leslie Marmon Silko, and Gloria Anzaldúa.

CHAPTER FOUR

"Don't Fence Me In"

Barbed Wire in the Western

Oh, give me land, lots of land under starry skies above
Don't fence me in
Let me ride through the wide open country that I love
Don't fence me in
. . . .
I want to ride to the ridge where the West commences
Gaze at the moon 'til I lose my senses
Can't look at hobbles and I can't stand fences
Don't fence me in

—Robert Fletcher and Cole Porter, "Don't Fence Me In"

In December 1883, the *Austin Weekly Statesman* announced what it called "a Texas craze":

> The most recent craze in crime is fence cutting. The first man who applied the nippers to a wire fence is unknown, but the thing, like magic, when once begun, spread from one end of Texas to the other. It is a Texas institution, inaugurated in 1883, and destined to occupy the widest range while civil authority is loosely exercised. It took like fire takes an untrodden prairie in dry weather, and, no matter where it began, it has swept with devastating effect over Texas. It is getting just as natural now to cut a barbed wire as it is to want to eat. It is only necessary for one to look at a barbed wire for the craze to seize him. A peaceable and lawabiding citizen, going down Pecan street, has his fingers begin to work like a pair of nippers as he passes those stacks of wire at Firegaught's. He would just like to try a cut, and were he in the country, beside one of those long strings of wire fences, he would speedily be a fence cutter. This craze is nothing more nor less than the

> presence for a season of Satan in the confines of Texas. The spirit of the devil has taken hold of people, and the fence is desecrated around the homes of the dead as well as around the ranges of the long-horns.[1]

Foreshadowed by this journalist's lively prose, barbed wire swept through literature as it did across the dry Texas prairie. A poet and novelist with close ties to Texas journalism, Mollie E. Moore Davis, wrote the first novel to take up the role of barbed wire in the American West. Published in 1899, *The Wire Cutters* was written against the historical background of open range, early barbed wire enclosures, and the ensuing fence cutting.[2]

After a romantic, slightly lurid, sixty-two-page prelude set in the American South, protagonist Leroy Hilliard heads to West Texas, where he has bought a ranch sight unseen. He learns quickly, works hard, and develops considerable standing in the community. When dead cattle are found outside barbed wire stretched around communal watering holes, people turn to him: "If anybody can explain the bob-wire contraption, Leroy Hilliard can do it." Investigating the "swiftly spreading evil" in adjoining counties, Hilliard finds that "public roads and familiar byways were closed up; many small freeholds . . . were fenced in. . . . Springs, ponds, and water-holes were ringed about with the formidable wire. . . . Thousands of cattle had already perished." He leads some of the men to inspect the new fences, and they discover barbed wire strung across a county road to block access to a water hole. They cut the wire and go looking for the fence builders. When they find them, Hilliard steps forward and with more formality than is usual in West Texas, announces: "I have the honor to inform you that your fences have been cut wherever they encroached upon property belonging to other people—except this one. And I now propose to cut this." He does so, and while the fence builders shoot at him, "the last strand of wire snapped in the sharp nippers, sprang apart, and curled up with a whizzing sound."[3]

The next day, Hilliard rides the same road and finds a man entangled in the cut wire:

> His face, lacerated by the barbed wire dragged across it, had bled profusely. . . . In the barbed wire which coiled around his neck and shoulders like a steel spring and trailed off to where the horse was standing, Hilliard read the story. . . . The rider, taken unaware, had pitched forward and fallen heavily into the mass of quivering wire. . . . It took Hilliard some time, working with infinite patience and

> gentleness, to extricate him from the jagged and dangerous wire. . . . This done, he drew the fatal wire clear of the road.[4]

The story Hilliard "reads" here and elsewhere is an emotionally charged story in which the material protagonist is "creeping like a steel centipede across the prairies."[5] Barbed wire is a vivid and vibrant force, a lacerating, formidable, quivering, jagged, dangerous, and evil wire. The vicious fence is so tightly strung that it whizzes and curls with barbed fury when cut. Davis endows barbed wire with a life force as threatening as any inanimate but inexplicably animate horror-movie demon. And like apocalyptic Zombies, *The Fog*, and *The Blob*, the novel's barbed wire fences are spreading swiftly with dire consequences.

At this point, Davis has established that barbed wire is a dangerous tool and that it can be used to take over public land. Because she knows Texas, and because she is writing historical fiction, she is also aware that barbed wire can be a useful tool. Her supremely reasonable Hilliard constructs a small dam on his own land and protects the water it slowly collects with a barbed wire fence. A wire cutter when need be, he is also a wire stretcher when it suits his purpose. His study of property and enclosure law leads him eventually to a career as a lawyer.

At a gathering of armed men remonstrating against "the barbed-wire outrage," Hilliard's antagonist and half brother Allan Deerford, the man found lacerated by barbed wire, delivers a rabble-rousing speech: "The barbed-wire business . . . is a damnable outrage. . . . By what right does the bloated capitalist of the far north and the far east . . . dare to invade the sacred homes and the inviolable birthright of the poor, perhaps, but free and manly Texan?"[6] The speech is straight from Texas newspapers during the wire cutting frenzy. Krell cites two examples that exhibit the flavor and substance of Deerford's argument. The first is a racist, anticapitalist, and religiously justified note found in Coleman County and published by the *Fort Worth Daily Gazette* on November 7, 1884:

> Down with monopolies, they can't exist in Texas and especially in Coleman County; away with your foreign capitalists; the range and soil of Texas belong to the heroes of the South; no monopolies and don't tax us to school the nigger; Give us homes as God intended, and not gates to churches and towns and schools and above all give us water for our stock.[7]

The second, from the *Galveston Daily News*, December 13, 1885, explains the fence cutters on populist and "communistic" grounds: "Fence-cutting never would have become so great and destructive if it had not met with such popular sentiment. . . . It found its way into the fireside of every home, and the grievances of the lawless element of the communistic fence-cutters was [*sic*] held up in glowing colours."[8]

These are not isolated views. Our survey of Texas newspapers found plenty of fodder for Deerfield's rhetoric. In a long piece on September 27, 1883, the *Austin Weekly Statesman* expressed what it saw as the prevailing public opinion:

> It is sufficient, perhaps, to say that for many years past, federal legislation, and to an extent that of many of the states, has forsaken the old and safe paths of prudence and patriotism, and has been so managed as to enable gigantic monopolies and corporations, and private individuals, to accumulate dangerously large fortunes in connection with great territories in land and the control of large numbers of men, dependent upon and receiving their orders from these wealthy land, stock, railway and bond owners. . . . Barbed wire fences have been set up all over the country, the parties so doing, in some instances, violating the law of the state by closing roads and obstructing travel, and overlooking private rights, as well as public, by fencing in great territories heretofore open, and much of which is not owned by those fencing it, and preventing the passage of the public, or access to water, by failing or refusing to put gates up, or by placing them where they can be utilized. . . . Public opinion has, for a time at least, given encouragement to the unlawful acts of those who have been cutting wire fences.[9]

Incited by such ideas, Deerford and his followers cut fences far and wide. They respond to Hilliard's opposition to their indiscriminate war on barbed wire by cutting his fence and destroying his dam. The wire cutters are ambushed one night and in the ensuing fight kill two of their attackers.

Violent responses to the fence cutters were predicted historically, as attested by this threatening statement quoted by the *Fort Worth Daily Gazette* from the *Texas Live Stock Journal*:

> It is not to be expected that such men will allow their property to be handled by outlaws be they many or few, in stealing live stock or destroying their fences. . . . Stockmen should proceed against such

men by the same means as they would for the capture of a gang of cow thieves and to act on such a basis. If the law does not furnish stockmen protection for their property, no class of men require less instruction as to how they should do it themselves.[10]

Anticipating just such a fate, Deerford flees the state after killing one of his own young followers, leaving the other wire cutters to be arrested. Despite his disagreement with their actions, Hilliard defends the young men in court, and his impassioned plea wins their release. Justice, it seems, is well tempered by mercy when the crime is provoked by a lacerating, jagged, and capitalist steel centipede and committed by well-meaning young men. In the end, Hilliard works with Texas lawmakers to enact a bill governing fencing with barbed wire, "forcing the State legislation on the subject which finally restored harmony."[11]

Texas newspapers provide historical context for Hilliard's reasoned attempt to bring order to the situation. A section from the long piece in the *Austin Weekly Statesman* quoted previously is representative:

> It is to be hoped that a healthy public sentiment will be aroused all over the state, and by the power of which all parties to this question may be induced to regard the rights of others and of the public as well as their own. . . . Legislative authority has been and will be invoked in this matter, but a modification of the present laws on these subjects, or an increase of penalty for these offences [*sic*], however desirable, will prove of slight value unless preceded by prompt and positive action of the people themselves. . . . Let the large pasture men deal justly with the community and with the smaller stock men, by having gates at all points needed for public convenience, and by giving access to water. . . . Let all unite, then, in a determination that fence cutting must cease, and that the full power of the law shall be brought to bear against parties guilty of this offense, and this difficult and dangerous question will be practically and peaceably determined in a short time."[12]

This would fit nicely in the final chapter of Davis's historical novel, although Davis chooses reconciliation rather than the full power of the law.

The Wire Cutters touches on several of the themes that will make barbed wire a staple of Western novels over the next century. Barbed wire is an ambivalent tool of control, despised or embraced by cattlemen depending on their aims. There are good wire cutters and bad wire cutters, depending on what has been fenced in or out. And in the end, although it is coiled, quivering, dangerous, and fatal, although there are few kind words to describe the fence that works because it causes pain, barbed wire becomes the standard fence of the American West and a favorite source of conflict in the Western novel.

Historian Frederick Jackson Turner announced the closing of the American frontier in an 1893 essay, three years after the federal census noted there were no more vast tracts of unsettled land in the West.[13] Turner's frontier had undergone decades of "civilizing" exploitation by waves of Euro-American settlers. Although his account glosses over the violent encounters between competing residents of the western frontier, other storytellers made their mark on American memory by emphasizing the violent clash of frontier cultures. Wild West shows, for instance, glamorized and exaggerated individual and group encounters with "savage" natives. Such versions of the frontier experience created a mythical West Americans clung to as the nineteenth century passed into the next century and beyond.

Richard Slotkin argues that the imaginative stories and fables arising from this frontier West helped people make sense of their experiences and gave meaning to their history. American writers constructed a "uniquely American mythology" as they attempted to "reshape the character of their people or to justify some preconceived or inherited notion of American uniqueness." Slotkin defines this "myth-consciousness" as the "continual preoccupation with the necessity of defining or creating a national identity." The frontier myth centered on the idea that America was "a wide-open land of unlimited opportunity for the strong, ambitious, self-reliant individual to thrust his way to the top," a place where these individuals could "regenerate their fortunes, their spirits." Slotkin argues further that "the means to that regeneration ultimately became the means of violence, and the myth of regeneration through violence became the structuring metaphor of the American experience."[14] Violence came in many forms, including the use of barbed wire to control contested resources and the cutting of fences to reestablish a "wide-open land."

In an ironic twist, individuals like Daniel Boone, Wild Bill Hickok, and other Euro-American pathfinders, Indian fighters, trappers, and hunters in the advance guard of eastern "civilization," were themselves dispossessed by more "civilized" settlers. As Turner's waves of civilization swept the frontier, the railroad, telegraph, plow, and barbed wire enabled new levels of conquest and exploitation, hastening the end of the conditions that gave rise to the myth of the West. The myth itself, however, endured, flourishing in popular consciousness. Visual artists such as George Catlin, Timothy O'Sullivan, Frederic Remington, Thomas Moran, and Charlie Russell joined a host of writers, including Owen Wister, Zane Grey, and Louis L'Amour, and filmmakers such as King Vidor, John Ford, and Sergio Leone, to depict an American western experience that occupied a space somewhere between myth and reality. Barbed wire, as one might expect, is an important motif in many of the novels and films that developed and perpetuated the myth of the West.

Published just three years after *The Wire Cutters*, Owen Wister's *The Virginian* (1902) inaugurated a new genre: the Western.[15] Where Davis's novel placed barbed wire at the historical center of a legal conflict resolved by a young man whose hallmark is thoughtfulness, the conflict in *The Virginian* rises to the level of myth with a hero who embodies both wilderness and civilization. The young, eastern narrator is infatuated with the mythical cowboy (as was Wister himself). He repeatedly describes the Virginian as tigerlike, marveling at the untamed aspects of his character.[16] The wild and the civilized are manifest in a kind of recurring dialectic: the Virginian becomes a ranch foreman; he takes part in a grim lynching of rustlers; he expands his education; he woos and wins the eastern woman who has come to teach school in Wyoming; he defies the wishes of his fiancée and shoots his antagonist; and finally he travels east with his new bride to meet her thoroughly (perhaps too thoroughly) civilized family.

Readers of the novel often note the Virginian's ability to act as a bridge between a savage wilderness and encroaching civilization. Jennifer Tuttle, to cite but one, argues that at the very moment when Native Americans are facing eradication at the hands of what Wister saw as racially superior white men, and in a novel marked by the virtual absence of Indians, Wister appropriates the supposed savagery of the Native American to enhance the Virginian's manliness.[17] This is reminiscent of the case noted previously in which the image of a fictional "Waukegan Chief" was used to sell a brand of barbed wire produced in Waukegan, Illinois.

Like the Virginian, the barbed wire fence has a double nature in the novel. It is a savage fence that enables civilization in a harsh environment:

> They came upon the schoolhouse, roofed and ready for the first native Wyoming crop. It symbolized the dawn of a neighborhood, and it brought a change into the wilderness air. The feel of it struck cold upon

> the free spirits of the cow-punchers and they told each other that, what with women and children and wire fences, this country would not long be a country for men.[18]

Women, children, and wire fences are a triple threat to the free-spirited cowboys. That barbed wire is included in the domestic trinity is due, in part, to the efficacy of a quarter century of advertising that promoted the vicious barbed wire as an enabler of domestic order and safety. And if barbed wire is women's equivalent, then it makes them, in this somewhat misogynistic novel, barbed, dangerous, and controlling, which is to say, civilizing.

The novel describes borders of several kinds, the most important of which stand between civilization and wilderness: "The long gray line of fence . . . stretched along the waste ground and added desolation to the plain. . . . The last gate was shut by the Virginian, who looked back at the pleasant trees of the ranch, and then followed on in single file across the alkali of No Man's Land."[19] Action in the novel alternates between inhabited (fenced) and uninhabited (unfenced) country. Both the lynching that is the dark heart of the novel and the pursuit of two rustlers are enacted in wilderness that the narrator describes as overwhelming: "The tall peaks rose like spires around it, magnificent and bare in the last of the sun. . . . Our bleak, crumbled rim ran like a rampart between the towering tops, . . . very wide in some parts, and in some shrinking to a scanty foothold."[20] Immanuel Kant wrote about human interaction with threatening nature as an experience of what he called the sublime: "The might of nature forces upon us the recognition of our physical helplessness as beings of nature, but at the same time reveals a faculty of judging ourselves as independent of nature . . . that is the foundation of a self-preservation."[21] Wister's novel repeatedly constructs the dual experience with nature that constitutes the sublime. Outlaw threats to civilization constitute a similar duality in the novel. The Virginian is reluctant to be governed by the constraints of civilization—boundaries and bonds symbolized by barbed wire fences and schools and women. Reluctant, but not entirely unwilling. The tension is on display as he rides to his much-anticipated marriage through more civilized (and thus less sublime) country: "In the virgin wilderness the scars of new-scraped water ditches began to appear, and the first wire fences. Next, they were passing cabins and occasional fields, the outposts of habitation. The free road became wholly imprisoned, running between unbroken stretches of barbed wire."[22] The ominous linkage of imprisoning barbed wire and marriage underscores the narrator's (and Wister's) ambivalence toward marriage. The Virginian is leaving the "virgin wilderness" to consummate his marriage with civilization.[23]

After the final, fatal encounter with Trampas through which the Virginian once again proves his manly courage and skill, and following their marriage the next day, the Virginian and his bride, Molly, travel out of the town's barbed wire prison: "They passed through the gates of the foot-hills. . . . The outstretching fences and the widely trodden dust were no more. . . . The ploughed and planted country . . . lay in another world from this where they rode now." The beautiful setting of their honeymoon is not the sublimely threatening mountains, but neither is it the constricted setting of the town. Nonetheless, civilization and its fences are inevitable, as the Virginian explains to Molly's great-aunt when they travel east to meet Molly's family: "We are getting ready for the change—some of us are. . . . When the natural pasture is eaten off . . . we'll have big pastures fenced, and hay and shelter ready for winter."[24] Like Davis's Leroy Hilliard, the wise Virginian eventually makes peace with a fence as enabling as it is inhibiting.

The Wire Cutters and *The Virginian* set the stage for the barbed wire Westerns that follow. Following Davis's plot driven by questions of enclosure law and after Wister's savage and civilized cowboy who becomes a prosperous builder of fences, writers of Westerns often feature conflicts related to barbed wire. Works with titles like *The Drift Fence*, one of Zane Grey's 1930s Arizona novels; Elmer Kelton's 1957 novel *Barbed Wire*; and Cameron Judd's 1981 novel *Devil Wire* are prime examples. A quick survey of these works; Louis L'Amour's *Kiowa Trail* (1964); and several Western films, including *Shane* (1953) and the *Man without a Star* (1955), reveal an obsession with the role barbed wire played in establishing a Western Identity.

Zane Grey's Arizona novels depict a feud between the Traft and Dunn families as ranchers begin to fence the open range for private use. The protagonist of *The Drift Fence*, Jim Traft, travels from Missouri to Arizona to become foreman on his uncle James Traft's large Diamond Ranch. Young Jim's first assignment is to build a hundred-mile barbed wire drift fence to keep cattle from drifting into the brakes. His uncle knows his fence will be unpopular, and he anticipates serious consequences:

> "You see, barbed-wire fence in this country is nothin' short of murder. An' these nesters an' homesteaders an' backwoodsmen will lose by it. An' as for cowboys, Lord! how they hate any kind of a fence! I reckon I used to. But I'm ahead of my day. I can see what is needed an' what

is comin'. So, Jim, you can trust me so far as the benefit to ranchers is concerned."

"Uncle, have you an actual right to fence the range?"

"Shore, but it's open to question. If it ever went to law I'd have to prove my contention. An' whoever sued me would have to show why he was bein' hurt. No honest cattleman will ever take such action. An' I'm shore doubtful about any of the homesteaders goin' so far." . . .

"I see. So far so good. How about your moral right?" . . .

"The question of moral right . . . is the most open to argument. My own conscience is clear on that. I *know* in the long run the range will benefit." . . .

"These honest ranchers will just stop appropriatin' calves they're not shore they own. . . . An' others will cut our fence an' go right ahead stealin'. That's where our trouble will come in."[25]

Young Jim nearly loses his life as fence cutting ensues and then escalates to kidnapping and killing, not unlike what actually materialized in Arizona in the Pleasant Valley War of the mid-1880s.[26] Here and in many of the Westerns that follow, barbed wire is at the center of historically based conflicts between defenders of free range and settlers and ranchers who want to fence land for private use. Formulaic for the most part, the genre novels and films are nonetheless interesting as they inscribe barbed wire into the myth of the West.

In Elmer Kelton's 1957 novel *Barbed Wire*, for instance, there is considerable discussion of the qualities of barbed wire. It is a boon to cattlemen: "They're using a lot of it down in South Texas. It's the answer for the stock men. . . . Country's too dry to grow hedges, and it'd bust the back end out of a bank to build a wood fence around a big pasture. But barbed wire is cheap. Most anyone can afford it." Barbed wire is also a destroyer of the open range: "It's always been an open range. It's been *our* range. Once a few of the farmers start, some of the ranchers will. In a couple of years they'll have the range cut up into a hundred pieces. We'll be fenced off from half of our water. The cattle won't be able to graze free with the rain and the grass." In Kelton's story, a young Doug Monahan is hired by ranchers to string barbed wire in South Texas. The fence stretching pits Monahan against cattle baron Captain Rinehart, an unscrupulous man who cuts fences and orders murders to forestall the closing of the open range. Kelton's story includes repeated discussions between characters about the economic problems raised by barbed wire: "It takes a lot of cowboys to keep a cow outfit running, the country wide open like it is. But you cut this land up into little pieces, it won't take near as many men to work it. A lot of those boys'll be out of a job, and they're smart enough to see that already."[27] Much is said about the economic promise of barbed wire as well:

> As long as the range is wide open, how can a man develop his own land, other people's cattle crowding into his water and onto his grass? . . . But you put a fence around him . . . he can build up the best cattle in the country. . . . He can fence the range hogs out. He can put a crimp in the cow thief. . . . With the fence the country will finally be permanent. It'll produce more and make a living for more people. There'll be more towns, and they'll be bigger ones. Sure barbed wire is going to hurt some people. But it'll help a lot more of them than it hurts.[28]

This theme of building a better herd through the use of barbed wire is echoed again in Kelton's 1969 *Shotgun Settlement*, in which one large rancher (Blair Bishop) uses barbed fencing to improve his herd: "Bishop's place was under fence now, and he was buying better bulls to upgrade his herd. Every time a C Bar bull came over, it meant some of Blair's cows were going to fetch up a scrubby calf."[29] In *Barbed Wire*, there is a final act of violence and then a brutal reckoning between Monahan and Rinehart's evil foreman. Because the plot is driven by the historical transfer of free range to fenced ranches and farms, the novel ends as expected with fence builder Monahan victorious and the other settlers grateful to be his new neighbor as he buys a ranch and offers his hand to a lovely young woman.

The plot of Cameron Judd's 1981 novel *Devil Wire* follows a similar arc, with barbed wire as the contested symbol of change. The novel is rich with barbed wire history; in fact, as he approaches his Montana destination, young Jim Hartford meets a salesman for Washburn & Moen who discourses at length on the virtues of his product. Hartford joins his brother Sam and other progressive Montana cattlemen who are fencing their range, mindful of the conflicts brought on by barbed wire in other places yet understanding the value of barbed wire in the West: "Settlers will keep on coming, barbed wire will keep on going up, and the old ways will keep on changing. That's for certain, and all the fence-cutting riders in the world won't be able to change it. If only people would learn." People have not, however, learned that lesson yet, and that drives the ensuing conflict. The area's biggest rancher hires men to terrorize the fence builders. They cut fences, shoot into cabins, and leave notes: "We have given our warning—the devil wire must go, else there will be blood on the land—no home, no wife, no child will escape—the land shall run red with fence-stringer blood."[30] Although he is struck by lightning and has to save his sweetheart by shooting a man with a keepsake Cheyenne bow and arrow, Hartford and friends win the day because they embrace the coming of a new West symbolized by the spread of barbed wire. And like all of the Westerns with barbed wire plots, Native Americans are present only in the form of useful relics. No need for them if the plot turns on who will control land already taken from the Natives.

In Jack Schaefer's 1949 novel *Shane*, the basis for the influential film with the same name, barbed wire is again the catalyst for conflict between Wyoming firstcomers and recent migrants, between large ranchers and small family-run outfits. The fences that begin to hem in Luke Fletcher's herd are embraced by Joe Starrett and other small ranchers: "The open range can't last forever. The fence lines are closing in. Running cattle in big lots is good business only for the top ranchers and it's really a poor business at that. Poor in terms of the resources going into it. Too much space for too little results." The conflict unfolds as it does in most Westerns. When the big rancher fails to convince the smaller spreads to sell and move out, a series of physical confrontations ensue, beginning with fence cutting. Into this battle rides Shane. Even the young narrator is aware that something is smoldering under the surface of this polite and constrained man, and he expresses that with a fencing metaphor: "He must have been riding many days, for he was full of news from towns along his back trail as far as Cheyenne and even Dodge City and others beyond I had never heard of before. But he had no news about himself. His past was fenced as tightly as our pasture."[31] Barbed wire is an efficient instrument of control, but when strung too tightly, the wire will snap and recoil dangerously. The physical act of fence cutting by Fletcher's men unleashes the tightly wound Shane, and all hell breaks loose.

George Stevens's 1953 film of the novel brings fences front and center as historical reality and as metaphor. As the story begins, Shane (Alan Ladd) rides through a valley, across a river, and up to a series of fences that visually dominates most of the scenes at the Starrett ranch: rail fences around the ranch house and corrals, a twine fence around a garden plot, and later barbed wire fences. The first interchange between Shane and Joe Starrett establishes the theme: "'Hope you don't mind me cutting through your place.' 'No, I guess not.' 'Heading North. Didn't expect to find any fences around here.'"[32] Settlement has taken place more rapidly than Shane had anticipated. Although he is headed north to put himself beyond whatever conflicts have formed his character, he is drawn to the Starrett family and resolves to stay with them for a time. Shane's first task as a hired hand is to drive into town to pick up a load of wire from the mercantile store, where spools of barbed wire are stored with coffee and ready-made clothing. In the course of the film, Shane and Starrett stretch and repair the barbed wire fences the homesteader relies on.

Although the various fences are signs of settlement, the borders those fences define and defend are somewhat tenuous. An elk dines casually in the garden, disregarding the string fence, and Ryker's men ride through the same garden fence to warn Starrett off his claim. Riders cut Starrett's neighbors' fences in the night and drive cattle into their wheat field. When Ryker (the Fletcher character from the novel) attempts to hire Starrett and buy his homestead, Starrett rejects the offer outright, and Ryker replies with his own litany of complaints:

> Look Starrett, when I come to this country, you weren't older than your boy. We had rough times. Me and other men that are mostly dead now. . . . We made this country. We found it, and we made it. Work, blood, and empty bellies. . . . We made a safe range out of this. . . . And then people move in. . . . They fence off my range and fence me off from water. Some of them like you plow ditches and take out irrigation water, and so the creek runs dry sometimes, and I've got to move my stock because of it. And you say we have no right to the range.[33]

The conflict between firstcomers and the next wave of settlers, between free-rangers and homesteaders, between large ranchers and small ranchers, is a matter of controlling limited resources. Consistently in these Westerns, the free-rangers resort to violence to achieve their aim while the homesteaders put up barbed wire fences to assert their more civilized control of the land. Shane, like the Virginian, is a dangerous man who has himself under tight control. When his friends are threatened by the violent free-range advocates, he protects them with skills they can only imagine. But where the Virginian acculturates himself to the new, fenced civilization, Shane moves on in an iconic gesture, headed for less settled country, riding off into the mythical sunset.

Matthew Costello's analysis of the film places it in the context of Cold War cultural changes and the "cultural ambiguities" that Americans faced "from changing gender roles, rise of suburban planned communities, and the growth of a bureaucratized economy."[34] Costello argues that the film's series of enclosures—encircling Teton Mountains and the various kinds of fences that define homestead claims and prevent movement of stock—represent the epic struggle between the common good and individual desires. Fences are a potent metaphor for constraints on human behavior.

A similar constellation of conflicts fueled by barbed wire can be found in King Vidor's 1955 film *Man without a Star*. The protagonist (Kirk Douglas as Dempsey Rae) is a roving cowboy whose past experience with barbed wire has left him deeply scarred, both physically and emotionally. In a scene with his young protégé Jeffrey Jimson, Rae recounts his love for the open range and explains why he left the Lone Star State,

> "What made you drift North? Isn't there any more grass in Texas?"
>
> "Lots of it kid. Tall sweet and it's real good. Stretching out from the Pecos to the Sabine. Then they rode in the wire. It was all open range as far as you can see. As far as a man could ride a horse or drive cattle there was nothing to stop you. Nothing to drag you down."
>
> "That's why you came up here?"
>
> "Yeah. Now the wire is catching up with me again. You know kid, you and I may just end up in Canada."[35]

Rae's new boss (Jeanne Crain as Reed Bowman) begins overstocking the open range at the expense of smaller outfits, and he decides it is time to pull up stakes and ride out to avoid the impending conflict. At this point in the film we learn Rae's brother was killed during a barbed wire conflict in Texas, and we see the vivid scars left on Rae's chest by barbed wire. Rae's nemesis from the Texas conflict, Steve Miles, rides in with a herd and is hired by Bowman to run her outfit, replacing Rae. The range war commences, dead bodies accumulate, barbed wire is cut, good cowhands leave or are killed, and Bowman's Texas gunmen rule the range.

Rae must now confront his own personal history with barbed wire. He joins a small outfit and begins to string the very wire that heralds the end of the cowboy on the free range. When Miles's men attack a fencing crew, Rae wraps the lone survivor in barbed wire and delivers him to the Bowman ranch as a warning against further fence cutting. The conflict ends with Rae beating Miles into a barbed wire fence, leaving him hanging on its barbs. Although the grateful small ranchers offer Rae a small ranch of his own, he refuses: "I just don't like barbed wire." The final scene shows Rae riding westward with barbed wire in the foreground, fleeing the silvery strands to seek a place where they are still absent, a lone figure exiting the physical and metaphorical constraints of the wire. The control barbed wire exerts is good for some people. Not for the mythic cowboy.

Louis L'Amour's 1964 *Kiowa Trail* features a rancher named Kate Lundy and her crew at the end of a cattle drive in Kansas. The gender of the boss is an early signal that this novel will deviate from the norm. Lundy's brother is murdered after crossing a strictly enforced border between the town's saloons and the civilized domiciles of the town's business elite. In retaliation she encircles the town with a barbed wire fence. "East of here . . . a train will be unloading barbed wire. I have leased the railroad land on both sides of the tracks, and I'm going to fence the town in. . . . Your town does not deserve to live, and it shall not."[36] The fence strangles the town's economic activities, forcing inhabitants to abandon their businesses and residences as the town dies. In L'Amour's novel, the generic expectation pitting fence-cutting free-range advocates against fence-building settlers is replaced by free-range cattle drivers who build their own fence, a border more effective than the one dictated and enforce by the townspeople. If the novel continued into the future, which it does not, barbed wire and the penetration of these territories by railroads would eventually end Lundy's cattle drives. Barbed wire drastically alters ranching forever in the real, if not in the imagined, West.

Proof, if it were otherwise needed, that barbed wire has become an internationally recognized staple of genre Westerns can be found in De Bevere and Goscinny's "Lucky Luke" comic book *Des barbelés sur la prairie / Barbed Wire on the Prairie*. In the French caricature, skinny vegetarian settlers fence their lettuce fields to protect them from the cattle of the gluttonous beef-eating residents of Cow Gulch. When the obese carnivores cut their fences, the settlers enlist the help of easygoing gunslinger Lucky Luke. The conflict ends when the settlers stretch a circle of barbed wire around the town—à la L'Amour's *Kiowa Trail*—a turnabout that forces the cattlemen to compromise with the vegetarians: "You're right. The West is big enough for everyone, our herds will respect your farms! . . . And we will replace the barbs with flowery hedges and fruit trees!"[37]

Wallace Stegner thoroughly disliked the mythic hero these Westerns evoke. In an essay titled "Variations on a Theme by Crèvecoeur," he describes the character:

> The Virginian is the standard American orphan, dislocated from family, church, and place of origin, with an uncertain past identified only by

> his nickname. With his knightly sense of honor and his capacity to outviolence the violent, he remains an irresistible model for romantic adolescents of any age, and he transfers readily from the cowboy setting to more modern ones. It is only a step from his "When you call me that, smile" to the remark made famous by a recent mayor of Carmel and by the fortieth president of the United States: "Go ahead, make my day."[38]

Stegner writes that he "would obviously like to bury [this mythic figure]. But I know I can't. He is a faster gun than I am. He is too attractive to the daydreaming imagination. It gets me nowhere to object to the self-righteous, limited, violent code that governs him." Why is this the case? Stegner thinks it is a matter of space: "Space, itself the product of incorrigible aridity and hence more or less permanent, continues to suggest unrestricted freedom, unlimited opportunity for testings and heroisms, a continuing need for self-reliance and physical competence."[39] Western space, of course, is structured by barbed wire fences.

Gretel Ehrlich continues Stegner's thoughts on fences, space, and the demand for unrestricted freedom:

> Fencing ultimately enforced boundaries, but barbed wire abrogated space. It was stretched across the beautiful valleys, into the mountains, over desert badlands, through buffalo grass. The "anything is possible" fever—the lure of any new place—was constricted. The integrity of the land as a geographical body, and the freedom to ride anywhere on it, were lost.
>
> I punched cows with a young man named Martin, who is the great-grandson of John Tisdale [who was shot in the back by an agent of cattle barons]. His inheritance is not the open land that Tisdale knew and prematurely lost but a rage against restraint.[40]

Erlich's description of the losses incurred through fencing and of the cowboy who needs fences for his cattle but who has a "rage against restraint" brings us out of the traditional Western and into the literary terrain of the New West.

Before we turn to the fiction of the New West, a final note about barbed wire and the Western. On a poster for director Martin Ritt's 1963 Western film *Hud*, Paul Newman's character, Hud Bannon, is described as "the man with the barbed wire soul!"[41] Wild, law evading, rapacious, and in the end desperately lonely, *Hud* is well deserving of the epithet. The shift here from actual barbed wire to barbed wire as metaphor draws on vicious aspects of the fence early advertisers skirted gingerly as they established meanings more conducive to widespread consumption.

When Davis wrote the first novel about barbed wire, *The Wire Cutters*, and Wister his influential *The Virginian*, they were well aware of the historical course of barbed wire, wire cutters, free-range cowboys, and the Old West. What they could not anticipate was the broad range of meanings that would be assigned to barbed wire in the New West. Early barbed wire meant what it was used for: control of the range and control of cattle. Opposition to it was stated in terms of free range and freedom from the shackles of civilization. But after the free range was gone and after the cowboy (and Indian) had been forced to submit to civilization and its discontents, barbed wire took on a new set of meanings.

CHAPTER FIVE

"Intimate Fences"

Barbed Wire in the New West

He could recall the intimate fences he'd made, taut wire and perfect corners.
—Annie Proulx, *Close Range*

Dismantling the myths of the Old West and exploring what they call the New West has been the task of historians such as Howard Lamar, Earl Pomeroy, Patricia Limerick, Donald Worster, Richard White, and William Cronon.[1] The New West is both rural and urban. It is diversified by race, class, gender, politics, religion, and region. It is profoundly dependent on shared and contested resources. Technology, much of it new but still accompanied by and often symbolized by barbed wire, is a major feature of the New West. The savage/civilized Virginian, taut-strung Shane, Dempsey Rae with his vivid barbed wire scars, and so many other mythical cowboys are ghosts of the past. Many of the characters who replace them still ride horses and stretch barbed wire and raise cattle, but they inhabit a more real and more complexly conflicted landscape. Much like the persona in Amy Clampitt's poem "The Woodlot," the lives of people in the New West are shaped in various ways by the barbed wire that surrounds them. In this chapter we examine the barbed wire motif in literary works by several writers of the New West, including Edward Abbey, Doug Foreman, Bill Haywood, Amy Irvine, Ellen Meloy, James Galvin, and Annie Proulx.

Edward Abbey's 1975 *The Monkey Wrench Gang* revives the art of wire cutting. Although the gesture is similar, the objective is quite different from the one common to genre Westerns. In this case the war is not between free-rangers and settlers; these fences are cut by radical environmentalists in defense of wilderness. After "monkeywrenching" what they identify as instruments of exploitation and destruction—a bulldozer, an expensive set of geophones, and a drilling rig are disabled—two members of the gang drive along

Utah's Kaiparowits Plateau and contemplate its future at the hands of oil, power, and coal companies and land developers. Down off the plateau, they encounter an additional creator of wasteland: a herd of cows. Incensed, they cut the fences that keep the cows on their intended grazing plot. While they cut, they talk (and the narrator intervenes to provide an environmental assessment of barbed wire):

> "You can't never go wrong cuttin' fence," Smith would say. "Especially sheep fence."
>
> (*Clunk*!) "But cow fence too. Any fence."
>
> "Who invented barbed wire anyhow?" Hayduke asked. (*Plunk*!)
>
> "It was a man named J. F. Glidden done it; took out his patent back in 1874."
>
> An immediate success, that barbwire. Now the antelope die by the thousands, the bighorn sheep perish by the hundreds every winter from Alberta down to Arizona, because fencing cuts off their escape from blizzard and drought. And coyotes too, and golden eagles, and peasant soldiers on the coils of concertina wire, victims of the same fat evil the wide world over, hang dead on the barbed and tetanous steel.
>
> "You can't never go wrong cuttin' fence," repeated Smith, warming to his task. (*Pling!*)
>
> "Always cut fence. That's the law west of the hundredth meridian. East of that don't matter none. Back there it's all lost anyhow. But west, cut fence."
>
> (*Plang!*)[2]

In this novel, the act of wire cutting becomes an environmental obligation (fig. 33). Like the strip mines and drilling rigs that destroy wilderness to get at whatever can be extracted, fences controlling cattle herds that lay waste to a delicate desert landscape must go. The "cowboys" of the New West—environmentalist wilderness lover, the polygamist river runner, and the Vietnam vet and explosives expert—are the new wire cutters.

Like all proverbial fence cutters, the members of the Monkey Wrench Gang eventually run afoul of the law and must pay the consequences. But the spirit of Abbey's war on technologies designed to control and exploit nature continued with the 1980 organization of Earth First! by Doug Foreman, Bill Haywood, and several of their comrades. Foreman and Haywood edited the book *Ecodefense*, which features a frontispiece image with a masked, battle-ready,

Figure 33. "You Can't Never Go Wrong Cuttin' Fence." Illustration by Robert Crumb. From Edward Abbey's *The Monkey Wrench Gang* (Salt Lake City, UT: Dream Garden Press, 1990). Copyright © Ken Sanders. All Rights Reserved.

superhero monkeywrencher in the breach of a freshly cut barbed wire fence (fig. 34). A toppled billboard behind him announces "PLUNDER ACRES."

In the eyes of Foreman, Haywood, Abbey, and others like them, the original technological sin committed by builders of railroads, telegraph lines, and barbed wire fences has been compounded many times over by the late twentieth century. Still, as evidenced in T. O. Hellenbach's comments in the second chapter of *Ecodefense*, barbed wire remains an identifiable symbol of the more general outrage: "Throughout most of the land, the Wolf has vanished, barbed wire rules, the natives have lost their sacred soil, and we are largely slaves to the industrial culture born in the coal-fired furnaces of Europe."[3] In this throwback to Thomas Jefferson's pastoral vision, the wolf has vanished and animals like coyotes need to be defended. The section of *Ecodefense* titled "Fence Cutting" responds to the killing of coyotes and suggests a Luddite response:

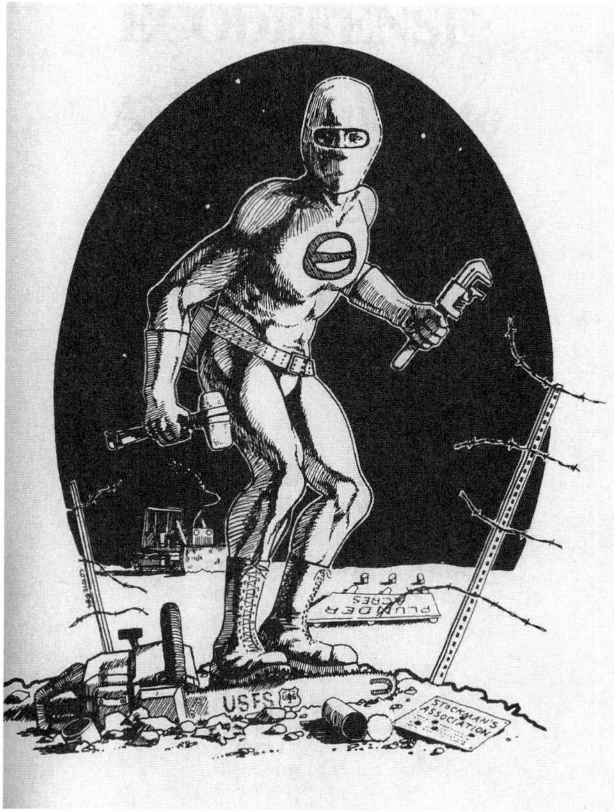

Figure 34. Ecoman. From *Ecodefense: A Field Guide to Monkey-wrenching*, 2nd ed., Dave Foreman and Bill Haywood, eds. (Tucson: Ned Ludd Books, 1987).

> One of the ugliest sights you are likely to encounter while traveling about the West is the all-too-familiar dead coyote dangling from a rancher's fence. Never are you presented with a better reason to start straddling the fence and start cutting it. The tactics used will vary depending on whether you're in your neighborhood, or just passing through. Your primary consideration must be to avoid being caught, since small town juries and judges in ranching country are not likely to sympathize with a goldang coyote lover.[4]

Coyotes interfere with ranch herds. Environmentalists open up the land by cutting the fences that make ranching possible. When either party is captured, the consequences can be severe.

The conjunction of coyote, barbed wire, environmentalist, and rural justice is portrayed in a poignant scene in Amy Irvine's book *Trespass: Living at the Edge of the Promised Land.* The author is working in southern Utah for the Southern Utah Wilderness Alliance, an environmental organization detested by most of her rural neighbors. Despite her heart-wrenching and sometimes hilarious attempts to become a member of the local community, as an environmentalist she is a trespasser. One day she and her partner encounter a coyote with which she instinctively identifies: "Strung across the fence is a dead coyote. Blood drips from a fresh wound, blooms like a red Oriental poppy in the snow. . . . We climb over the fence and then lift the coyote off the barbed wire."[5] Hanging dead animals on barbed wire fences is a curious practice, meant to warn, perhaps, or meant to display power, or as a supplication to Nature. During the dust bowl, "a popular method of producing rain consisted of killing a snake and 'hanging it belly-side up on a [barbed wire] fence.' In the great drought of the 1890's an all but despairing Bohemian farmer ruefully told a passerby: 'I've killed three snakes and hung them on the fence, and each time we got a sprinkle of rain. If I could find enough snakes we'd get plenty rain.'"[6]

A hay truck approaches, and Irvine watches the driver, thinking how integrally connected he is to the place and to the people of the place. She envies him even as she knows that he may be the one who killed and displayed the coyote. She is an outsider, a trespasser, but she has her role in this place as well: "For now, it is enough to know that the coyote is off his crucifix. And that, at least within this tree-shrouded, makeshift gravesite, concealed from his persecutors, there is a prayer of a chance that he might be allowed to remain in dignified repose, ears and all."[7] Who belongs and who does not? The barbed wire demarcates and enforces the difference while it goes about its more mundane business of keeping in the cows. Irvine understands the need to belong, to be inside the fence, to share common experience. Her grandparents were Mormon ranchers; this is her culture too. *And* she is an environmentalist and identifies with the creature that crossed the fence and was made to pay with its life.

"*Open range*," she writes later in the book, "I love the phrase. The wide-open, no-limits kind of feeling it inspires."[8] But open range in the New West also means access for increasingly powerful off-road vehicles, for invaders of a new species. In Moab, Utah, Irvine befriends two young rock climbers who have bought a section of land at auction. Unfortunately, their land has a corner used by off-roaders to access public land. The new owners fence off the route with barbed wire and post a "No Trespassing" sign. The fence is torn down, and the perpetrators leave a noose and a hand-painted sign reading "Tree Huggers Die." The owners put up a second fence, which is dismantled by a San Juan County's sheriff's deputy. You are fencing a public right-of-way, he tells them, and adds that he will arrest them if they obstruct it again.[9] Control is the issue, and barbed wire is the means. Ranchers control their herds with barbed wire. Rural residents protect their property with barbed wire. Trespassers are punished on barbed wire. The merits and consequences of barbed wire control depend on perspective. Environmentalist monkeywrenchers cut fences in defense of wilderness, and as predicted in *Ecodefense*, "small town juries and judges in ranching country are not likely to sympathize with a goldang coyote lover." Advocates for open public lands for their off-road vehicles also cut fences and may find support in their communities.

In the ongoing contest between the federal government and people who want government out of their lives—called the Sagebrush Rebellion in earlier days—barbed wire continues to play a role. In 2016, at the Malheur National Wildlife Refuge in Oregon, fences were cut once again. Angered by federal ownership and control of land throughout the West, and specifically enraged by a new prison sentence given to two ranchers who had been ignoring government restrictions on where they could graze their cattle, armed protesters occupied the refuge and in the course of the occupation cut the barbed wire fence that kept cattle out of the refuge. Abbey's monkeywrenchers cut fence to let cattle out of sensitive range; these fence cutters wanted to let the cows back in. During the occupation, barbed wire was also invoked as a threat. Zach Cartwright's piece in the online journal *U.S. Uncut* bore the headline "Ranchers at Center of Oregon Standoff Threatened to Wrap Official's Son in Barbed Wire and Drown Him." Forrest Cameron, the *Uncut* piece reports, had been manager of the refuge from 1989 to 1999. He said that the ranchers Dwight and Steven Hammond had made repeated threats against him and his family: "My wife would take these phone calls—it was terribly vulgar language. They said they were going to wrap my son in barbed wire and throw him down a well. They said they knew exactly which rooms my kids slept in, in Burns [Oregon]."[10] Where control is exerted, especially in the American West, barbed wire is often the instrument. Given the tools of their experience, ranchers might have threatened instead to "hog-tie" the son, or to hobble him, or even to gun him down. But barbed wire was their threat of choice.

The Brave Cowboy, an Edward Abbey novel predating *The Monkey Wrench Gang* by nearly two decades, sets up the conflicts related to technology and barbed wire in somewhat different terms. Jack Burns, the title character, leaves the open desert to help his friend Paul Bondi, who is in jail for not registering for the draft. Access to Duke City (thinly veiled Albuquerque) is difficult. The first (and easiest) barrier confronting Burns and his horse is a barbed wire fence:

> Gleaming new wire stretched with vibrant tautness between steel stakes driven into the sand and rock, reinforced between stakes with wire staves. The man looked for a gate but could see only the fence itself extended north and south to a pair of vanishing points, an unbroken thin stiff line of geometric exactitude scored with a bizarre, mechanical precision over the face of the rolling earth. He dismounted, taking a pair of fencing pliers from one of the saddlebags, and pushed his way through banked-up tumbleweeds to the fence. He cut the wire—the twisted steel resisting the bite of the pliers for a moment, then yielding with a soft sudden grunt to spring apart in coiled tension, touching the ground only lightly with its barbed points—and returned to the mare, remounted, and rode through the opening, followed by a few stirring tumbleweeds.[11]

We have seen this "thin stiff line of geometric exactitude" that violates (or simply marks) the earth in the two poems we examined in the introduction to part 2. And the grunting wire that springs apart in coiled tension echoes the personification of barbed wire in *The Wire Cutters*. The busy highway is a more difficult obstacle: "the slippery pavement and the dense moving wall of steel and hard rubber. There was no possibility of outflanking this barricade."[12] They do finally get across, although another busy highway will exact its claim in the end. Before then, however, Burns (with files hidden in his boots) gets himself arrested so he can help his principled philosopher friend break out. The successive barriers he overcomes to get from the desert into the city and then into the jail are traversed in reverse when Burns breaks back out of the jail (Bondi will not follow him, law-abiding lawbreaker that he is) and leads the sheriff and the US Air Force on a perilous chase. Before Burns leaves the jail, Bondi asks about the government takeover of Burns's grandfather's ranch and whether he has ever returned. Burns reports that "the signs say 'Danger—*Peligroso*—Keep Out—Military Reserve—USAF.' Fence around the whole area and every gate padlocked. I had to cut wire to get in. Nothin but a few lizards. . . . A good place to set off atom bombs.'"[13] This is the Trinity site, the evil womb of nuclear war, and the nuclear arsenal is protected by barbed wire fences.

Art Hinton, the toilet-hauling truck driver who will eventually run over Jack Burns and his horse in the novel's culminating scene, is Burns's antithesis. Burns is as tough and able as any western cowboy, not yet a casualty of modern technology, but "brave" as in the title because it requires courage to make his way as a cowboy in the New West. Hinton is old beyond his years, sick and tired and hopped up on Dexedrine and coffee, an ulcered casualty of modern technology. The truck driver dreams of leaving his debilitating job, thoughts that grow especially vivid while he walks through a screaming wind outside a truck stop in Amarillo, Texas: "He staggered back, surprised and laughing, clutching at his cap. Between here and the North Pole, he recalled, nothing but *bob*wire."[14] When Hinton contemplates free and open spaces beyond the road that imprisons him, fences paradoxically symbolize open country—the country of ranches and cowboys and horses and health. Things have changed in the West. Barbed wire fences hinder free movement, true, but they also signal a rural, ranching landscape in which technology is somewhat limited. Highways, however, are lethal barriers for mounted cowboys, and the air force base from which planes scramble to find the brave cowboy and the Trinity site that spawned the atomic bomb are threats of a different magnitude altogether. In the New West, the conflicts symbolized by barbed wire are no longer arguments over free range for cattle. Besides making ranching possible, barbed wire fences mark and control borders surrounding the unthinkable.

Ellen Meloy's *The Last Cheater's Waltz* also sees connections between barbed wire and nuclear weapons, even while accepting the fence's rural utility. Because she is a westerner with property, she needs barbed wire to keep cattle out of her place:

> We constructed sturdy corners and pounded steel posts into sand and hard-packed clay. We strung two strands of barbed wire across the top and hog fence, or woven wire, along the bottom, ending each day looking like pincushions with cowboy tans. Our act was quintessentially western and adaptive. It established "control" over a parcel of ground in a region whose innate aridity and lack of wood for fencing materials required a vernacular technology to define boundaries: thin strands of Bessemer steel woven around vicious barbs.[15]

So far so good. But later, when Meloy finds a fence post that had not been set deep enough and looks around for a rock to pound it in more securely, she finds herself holding a piece of carnotite, the raw ore for uranium, and she screams.[16] The barbed wire fence and the uranium are not accidental associates in this narrative. Searching for claret cup cactus in bloom, Meloy drives through a section of the Navajo reservation and contrasts her own Anglo systems for marking property with what she knows of Native American practice:

> An intricate, largely invisible system determines customary land rights among the Navajo. . . . They recognize land tenure but not possession since, in the end, the land belongs to no one. Without the Anglo iconography of private property—survey pins, signs, barbed wire, gates, helicopter surveillance, remote sensory devices, half-starved Tibetan mastiffs—outsiders presume that Indian land is open range and that they can wander over any of all of it as they please.[17]

Meloy visits Los Alamos, a place that during the Manhattan Project had "a fence enclosing the town, guarded by military police." She contemplates a photographic exhibition of mushroom clouds she calls "wargasms." Searching again for claret cup cacti, she happens on abandoned mines where uranium was extracted, including a "water hole enclosed within the flimsy fence [that] likely served as a leach pond. . . . We hear that certain mine wastes are so harmless, we can *eat* the stuff; then they pack the stuff into million-dollar megabunkers surrounded by chain-link fence and razor wire."[18] Like Abbey, Meloy sees barbed wire as a somewhat innocuous ranching tool and at the same time a device with ties to technologies as destructive as the atomic bomb. Enclosing and controlling with barbed wire and releasing enormous energy by splitting atoms are polar opposites in physical terms, but in the minds of these authors it makes sense to see them as part and parcel of the same destructive enterprise.

A 1999 novel by James Galvin with the evocative title *Fencing the Sky* reimagines Abbey's *Brave Cowboy* in the context of a hard-scrabble ranching culture on the Wyoming-Colorado border. In this version of the American West, barbed wire is the saving grace of ranchers: it protects fields from incursions, limits the range of cows, ensures better care of livestock, and creates Robert Frost's proverbial good neighbors. That barbed wire is also somewhat dangerous is a simple matter of fact. Galvin catalogues the ubiquity and complexity of barbed wire in the landscape and lives of his characters, beginning with a

1904 incident when cattlemen attacked fence-building sheepmen and wired their hands together with their own "bob wire." There was the gut-shot buck that had jumped a barbed wire fence ("A six-foot length of intestine adorned the wires") and more recent trouble caused by bad fences ("Mike cut them out and put them through a fence that was barely there, the kind of loose and leaning barbed wire affair that cows have no trouble crawling through but that you don't take a horse anywhere near").[19] For Galvin, ranching is not the foreign element in this vast landscape; foreign is the land speculator who breaks up large ranches into forty-acre ranchettes for naïve believers in the mythic West, buyers lured by grand vistas and by visions of unfenced, cattle-free land. Galvin's old-timers know that enough forty-acre ranchettes inhabited short term by people whose fantasies are impotent substitutes for coping skills will mean the death of the West they value. The conflicting visions lead to trouble: "Most of the worst friction was about cattle. The new people seemed to bear a curious grudge against cattle, though they weren't so far as Ad could tell, vegetarians or environmentalists. The new people wouldn't build fences or fix existing ones, and they hated the law that said they had to fence cattle *out*."[20]

Fencing the Sky is an explicit response to Abbey's *The Brave Cowboy*. Like Abbey's Jack Burns, Galvin's Mike Arans is a transplant from the east and a former radical (Burns an anarchist, Arans a member of the Students for a Democratic Society). Like Burns, Arans is an honest-to-god cowboy:

> This was the life he had determined to lead. Hell or high water, he'd see this out. He'd made himself some promises twenty years ago, even before he'd got done with saving the world from capitalism and corporate rapacity. He would himself become a Utopian figure, an honest laborer in the fields, skilled torchbearer of horse culture in the western mountains and plains. In other words he wanted to be a cowboy when he grew up.[21]

That works fine for Abbey's and Galvin's characters until they run afoul of the law (Burns most dramatically when he breaks out of jail, Arans when he ropes and accidentally kills a man who is chasing his cows on an ATV). Both characters flee from Indian trackers and threatening helicopters to the novels' respective climaxes. Arans even thinks of himself in terms of Abbey's title: "The brave cowboy. The honest outlaw." Although both cowboys share a love of the work and of the land, Arans diverges from Burns's pattern in that he is married and settled, the owner of a ranch that requires building and maintaining fences. Even as he flees from the law, Arans holds true to the fence law of the range—if you cut it, you fix it: "Then he cut the fence, rode through, and took twenty minutes to stretch and splice the wires behind him. It was light. No use letting out stock and, worse, sending a telegram about his passage."[22] He crosses a fenced highway less traveled than the one where Burns meets his end, and once again cuts the fence with the skill of a fence builder. The barbed wire responds like a force of nature:

> He waited for dark, then rode down an open sagebrush slope to the highway department fence. He looped the reins over a post and walked two posts away to cut the wire. The four strands were tied by vertical twists, so he started at the bottom, and when he cut the top wire the fence rolled open like a wave. . . .
>
> He led Potatoes through and tied him to the other side of the same post. He took a coil of smooth wire from his saddlebags and spliced the wires behind him. . . . He couldn't make it look perfect, lacking a wire stretcher, but he did the best he could with what he had.[23]

For Galvin's character, maintaining fences is a means to preserve his way of life, and he splices the cut wire even as he flees from the lawmen who will end his life as a cowboy. That he does so in order to disguise his passage is almost beside the point. Arans knows barbed wire, uses barbed wire, respects barbed wire. He can make it roll open like a wave. When his flight and his life come to an end (although the ending is intriguingly ambiguous), the cowboy who lives with barbed wire steps into a simile: "The river was like a coil of concertina wire. Potatoes stood ankle deep at the edge of a rare eddy by the bank, which slipped off precipitously and married the white violence."[24]

Galvin's novel *The Meadow*, likewise set on an isolated ranch, begins with a scene in which a rancher describes what he sees from his window:

> Yesterday sixteen elk streaked across the hillside above the meadow. Lyle could easily imagine what they had done to the fence where it runs under deep drifts on the east side. They walked through it, not even feeling the barbs through their winter coats. They dragged broken wire through the woods, strewing it like tinsel. He'd find the pieces in the spring like tendrils of steel briar growing along the ground. It doesn't make him angry anymore, as it did in the early years. He figures the elk have been crossing that section of timber to forage on the north

> side of Bull Mountain for a lot longer than there has been anyone here to build fence and get pissed off every time the elk tear it up. Now he splices the fence with baling wire, which is lighter, so it will break easier and always in the same place and not get dragged so much or pull out posts.[25]

The writer whose *Fencing the Sky* works as a counterweight to Abbey's *The Brave Cowboy* here finds a solution to conflicts between the wild and the controlled, between the cows that must be kept in and the elk that must be allowed their freedom. It is a pragmatic compromise that might well appeal to ranchers and conservationists alike.

Mary O'Hara's much-loved novel *My Friend Flicka* (1940), like Galvin's novels, is less interested in the standard conflicts between barbed wire and free range than it is on the fence as a simple, if dangerous tool. Several scenes emphasize the potential for violence: "The whole web of wire was wrapped around the cow's udder. . . . The harder she pulled, the deeper the barbed wire embedded itself. . . . Ken . . . had been told by his father, *never let me catch you out without a pair of wire-cutters in your pants pocket.*" The novel's plot turns on a scene in which young Flicka, still wild despite Ken's attempts to tame her, tries to jump a high fence and is caught in the wire: "'Damn the wire!' cursed McLaughlin. 'If I could afford decent fences—' . . . she kicked and fought and thrashed until the wire was tightly wound and tangled about her, piercing and tearing her flesh and hide."[26] The story ends happily, but barbed wire remains a villain, tolerated only because the family cannot afford less dangerous fences.

Ubiquitous in rural Wyoming, the barbed wire in Annie Proulx's *Close Range* takes on a wide range of meanings. The men and women who populate Wyoming and these stories find that many aspects of their lives—insights and emotions and observations—make more sense when related metaphorically to barbed wire: "You can't have a fence with only one post"; "She had the lean of an old fence post"; "Well, excuse me all to hell. Cole, draw me another, will you? I'm goin a fight John I might as well have some a that liquid bobwire first"; "'You been a Mexico, Jack?' . . . He was cutting fence now, trespassing in the shoot-em zone."[27] There is hardly a story in the collection that does not feature barbed wire in one way or another, but three of the stories are especially interesting in their use of barbed wire in both real and metaphorical senses.

The story "All the Governors of Wyoming" brings a vegetarian, teetotaling spouter of Ed Abbey's manifestos onto a Wyoming ranch to cut fences. Wade Walls is his name, and he states his purpose with all the certainty of Abbey but with none of Abbey's mock-heroic masculine familiarity with the lives of actual westerners. Walls declares himself

> a monkey-wrencher, a hard man who would hammer a spike into a tree without hesitation. "You know what Abbey said about cows, don't you—'stinking, fly-covered, shit-smeared, disease-spreading brutes.' But it's not the way they are that matters, it's what they do to the land. They've wrecked the west, they're wrecking the world."[28]

The story raises questions about "divine order in a chaotic universe," about ranches threatened by rich developers "drinkin cappuccino while they watch the elk," and about conspiracy theories—"a powerful international group of men who want to control the ranchers and the farmers—to control the food supply of the world." Shy Hamp, an obsessive-compulsive "champion of nothing" obliged to leave college and take over the ranch when his parents died, wrestles with his own existential questions: "I can't *get* a holt on it. On anything." Hamp is tormented by the fact that he is a pedophile, just returned from two nights in a cabin with a "very young girl . . . a Shoshone girl from the reservation . . . a child." As a kind of penance, he turns to the environmental certainties of the simpering monkeywrencher: "This business with Wade Walls, which he had never thought out nor weighed beyond believing it a kind of good, served him as the balance column in the ledger of his own evil doings."[29]

A rancher discovers Hamp and Walls cutting his fences and shoots at them. Hamp is hit, knocked "on the ground, his good foot kicking against a steel post, the end of the cut wire rattling." The shooter shouts a barbed wire threat: "Son of a bitch, I have to come up there you'll come down wearin a bobwire necktie." Walls runs away and Hamp waits with an odd satisfaction and a muddled sense for the true penance that will come.[30] Relating fence cutting to moral depravity and ideological extremism and bolstering it with explicit quotations from Edward Abbey in the mouth of a self-righteous and hypocritical coward is a political provocation of sorts. Proulx tells her story from the Wyoming ranching perspective she has come to inhabit as skillfully as she once did the language and characters of Newfoundland in her novel *The Shipping News*, interested in the inner lives of her characters rather than in the environmental activism so important in Abbey's work.

Proulx's "The Half-Skinned Steer" gets its name from a tall tale told by an old man's raunchy girlfriend. One of the old man's sons, Mero, is twenty-three at the time; the other, Rollo, twenty. The tall tale is about a man known as Tin Head, named for the galvanized plate in his head. He is so down on his luck that when he kills a steer and skins half of it before dinner, he returns to find the half-skinned steer has wandered away. He finds it, looks into its red and hateful eyes, and knows he is done for. Mero remembers the story while driving back to Wyoming for Rollo's funeral. The story's first sentence lays out Mero's life in terms of wire—from tight wound to spooled out: "In the long unfurling of his life, from tight-wound kid hustler in a wool suit riding the train out of Cheyenne to geriatric limper in this spooled-out year, Mero had kicked down thoughts of the place where he began, a so-called ranch on strange ground at the south hinge of the Big Horns." The ranch had declined, and the boys' father had taken a job delivering mail. Mero and Rollo were left to do the ranching: "They kept mending fence . . . in the hope that sooner or later the old man would move to Ten Sleep with his woman and his bottle and they could, as had their grandmother Olive when Jacob Corn disappointed her, pull the place taut." Mending the fences and pulling the place taut had been a good plan for their forebears, but the repaired fences do not achieve the boys' aim, and "Mero wound up sixty years later as an octogenarian vegetarian widower pumping an Exercycle in the living room of a colonial house in Woolfoot, Massachusetts."[31] Now he is driving his Cadillac across the country for his brother's funeral.

It is a difficult and troubled drive, Mero's age working against him despite the Exercycle he uses like the once taut-pulled fences to combat the forces of entropy. Finally in Wyoming, approaching the ranch through a blizzard, "the shape of the ranch began to gather in his mind; he could recall the intimate fences he'd made, taut wire and perfect corners." The surprising evocation of "intimate fences" contrasts with everything else in Mero's decidedly nonintimate life. "Taut wire and perfect corners" should have set him up with a decent life, he muses, had he not felt forced to flee his father's Everclear-drinking girlfriend with her sexually threatening horsey haunches. Mero drives past the neighboring ranch, "the gate ghostly in the snow but still flying its wrought iron flag, unmarked by the injuries of weather, and the taut five-strand fences and dim shifting forms of cattle." He makes a wrong turn and finds himself stuck in the snow: "The map of the ranch in his memory was not as bright now. . . . The remembered gates collapsed, fences wavered, while the badland features swelled into massive prominence." He begins to walk, and "as he walked he noticed one from the herd inside the fence was keeping pace with him . . . the half-skinned steer's red eye had been watching for him all this time."[32]

As opposed to Shy Hamp, who cuts fence in a vaguely religious attempt to balance evil with supposed environmental goodness, Mero's taut and intimate fences are a denial of sexuality and death: the red and black Indian drawings of a vulva that so embarrass him when he is eleven or twelve; "the old man's girlfriend whom he imagined down on all fours, entered from behind and whinnying like a mare"; the "badland features swelled into massive prominence"; the half-skinned steer; and so on.[33] Proulx transfers the conflict between civilization and wilderness into the human psyche. Taut barbed wire fences are analogous, in this story, to Freud's version of a civilization established to control powerful instincts. And, as Freud theorized, Mero is patently discontented, a victim finally of psychological and biological entropy.[34]

In "Brokeback Mountain" (the source for Ang Lee's 2005 film), self-imposed fences like Mero's are thoroughly breached. The two cowboys tending a flock of sheep high on the mountain have no fences to deal with other than their initial inhibitions. When Ennis, who "ran full-throttle on all roads whether fence mending or money spending," finally joins Jack in his bedroll, they begin a sexual relationship that Ennis treasures for the rest of his life. Jack eventually moves to Texas and marries. Ennis marries as well. When they get together again, they wish circumstances had been different. Contemplating both past and present situations, Ennis remembers something his father showed him as a warning—an old cowboy living with another old cowboy who was finally killed with a tire iron, "drug him around by his dick until it pulled off."[35] Civilization exerts its control.

The two men meet now and again over the years. One year they argue about the infrequency of their meetings—Ennis is constrained by his work—and Jack mentions that he has been in Mexico. Ennis reacts with suspicion, cutting silences that are the fences enabling the difficult relationship: "'You been a Mexico, Jack?' Mexico was the place. He'd heard. He was cutting fence now, trespassing in the shoot-em zone." Jack shoots back angrily: "Measure the fuckin short leash you keep me on, then ask me about Mexico and then tell me you'll kill me for needin it." The men never get together again. Jack is killed in an accident, murdered with a tire iron, Ennis figures. Ennis drives to Jack's parents' ranch to ask for his ashes, which he wants to spread on Brokeback Mountain. Jack's father stares at him "with an angry, knowing expression," and "in the end the stud duck refused to let Jack's ashes go. 'Tell you what, we got a family plot and he's goin in it. . . . ' Bumping down the washboard road Ennis passed the country cemetery fenced with sagging sheep wire, a tiny fenced square on the welling prairie." Homosexuals must be controlled, confined within the barbed structures of the patriarchal western family. The sheep fence represents and enforces the law of enclosure in contrast to the wildness of the welling prairie and the mountain that "boiled with demonic energy, glazed with flickering broken-cloud light." Ennis and Jack, like Tin Head when he looks into the eye of the half-skinned steer, are done for: "This happen a other people?" Ennis asked about the desire between him and Jack, "What the hell do they do?" Jack answered: "It don't happen in Wyomin and if it does I don't know what they do, maybe go to Denver."[36]

The barbed wire of the New West, then, takes on meanings of various kinds. For environmentalists, barbed wire is a killer of wildlife, an enabler of grazing in sensitive landscapes, and a sibling of all the other nature-destroying technologies that culminate in nuclear weapons. For ranchers, barbed wire is an efficient tool that makes ranching possible in hard-scrabble environments and that contributes to their western identities. Proulx's "Brokeback Mountain" was published in the *New Yorker* in 1988. That same year, University of Wyoming student Matthew Shepard was beaten, tied to a buck-and-rail fence, and left to die outside Laramie. A Google search for Matthew Shepard and barbed wire, however, produces dozens of sources that extrapolate from news reports that he had been "tied to a fence" to the assumption that it must have been a barbed wire fence. Why this assumption? Because, perhaps, a barbed wire fence more closely approximates the homophobic violence than does a buck-and-rail fence. Because the barbed wire metaphors of the still-violent American West so avidly lend themselves to reification.

CHAPTER SIX

"The Thorny Fence"

Reifying the Religious Metaphor

It was a nun they say invented barbed wire.

—James Joyce, *Ulysses*

As an instrument of religious torture, according to Joyce's character, barbed wire was a godsend. Early references to barbed wire as the "thorny fence" made the analogy between Christ's punishing crown of thorns and barbed wire inevitable. Alan Krell traces several works of art in which a contemporary Christ figure is crowned by barbed wire, including a striking drawing of Nelson Mandela by Oliver Williams that shows the revolutionary leader with a double strand of barbed wire encircling his head.[1] Mandela was a prisoner on Robben Island at the time, and the crown of thorns, as Krell sees it, signals both captivity and sacrifice analogous to Christ's suffering for the good of humankind. In what follows, we discuss religious dimensions of the motif of barbed wire in a set of texts that includes John Steinbeck's *The Grapes of Wrath* and Flannery O'Connor's *Wise Blood*, and two short stories, Brian Evenson's "Contagion" and Jeff Mann's "The History of Barbed Wire." Cruel control is the specialty of barbed wire, and the barbed fence emerges in these works as an instrument of Christian asceticism, an object of religious veneration, and an enforcer and/or enabler of identity.

The Grapes of Wrath

An' I thought how some sisters took to beatin' theirselves with a three-foot shag of bobwire. An' I thought how maybe they liked to hurt themselves, an' maybe I liked to hurt myself.

—John Steinbeck, *The Grapes of Wrath*

It was these lines in an exchange between Jim Casy and Tom Joad about the conflicted workings of the "Holy Sperit" that first drew our attention to the barbed wire in Steinbeck's *The Grapes of Wrath*. This is a key scene in our reading of the novel, and we return to it later. First, however, a survey of fences in the novel.

Initially, barbed wire is simply part of the dust-bowl landscape: "In the morning the dust hung like fog. . . . It settled on the corn, piled up on the tops of the fence posts, piled up on the wires." Returning home after four years in prison, Tom Joad passes fences that reflect the men who erected them, hard-working men without the means to acquire better materials: "The right of way was fenced, two strands of barbed wire on willow poles. The poles were crooked and badly trimmed. Whenever a crotch came to the proper height the wire lay in it, and where there was no crotch the barbed wire was lashed to the post with rusty baling wire."[2] Although the Joad family and other farmers have been forced to become sharecroppers after banks foreclosed on their loans, the land still feels like their own, and the fence is a symbol of their pride:

> That there's our line. We didn't really need no fence there, but we had the wire, an' Pa kinda liked her there. Said it give him a feelin' that forty was forty. Wouldn't of had the fence if Uncle John didn' come drivin' in one night with six spools of wire in his wagon. He give 'em to Pa for a shoat. We never did know where he got that wire.[3]

Similarly, in his 1933 novel *To a God Unknown* Steinbeck signals the increasing prosperity of Joseph Wayne's California ranch by reporting that "the barbed-wire fences were edging out to encircle the land," and later, "Joseph was working tremendously in the prodigal spring. . . . Thomas and Joseph worked silently together, stringing the barbed-wire fences out around the land." The fences establish control over the landscape and mark it as the property of an ambitious and talented rancher, but when the extreme drought arrives, nature takes control and negates the human endeavor. The only use for the fences during the drought are to enforce property rights: "The moving herds filled the road and the farmers along the road were hostile. They patrolled their fence lines and shot any stock that broke through."[4]

In *The Grapes of Wrath*, Grampa and Muley so identify with the land they have worked that they refuse to leave the forty acres, even though their awkward and largely symbolic and probably stolen fences have no power in the world of foreclosures and profit margins. The two strands on willow poles are ineffectual defenders against the snub-nosed momentum of the capitalists' Caterpillar tractors: "across the country, through fences, through dooryards, in and out of gullies in straight lines. They did not run on the ground, but on their own roadbeds. They ignored hills and gulches, water courses, fences, houses."[5] The fences erected to mark forty acres of hard-won property are simply swept away.

The Joads' barbed wire fences are harmless reminders of their failed dreams. In the comprehensive economy of Steinbeck's socialist novel, however, other fences exist that are less benign. Jim Casy, for instance, thinks about changes going on in the country, about the economic conditions that are driving thousands of families west, and about an inevitable broad uprising against the responsible system, and he frames this complex set of issues in terms of fences. His forward-looking responses contrast with Tom's more pragmatic approach, although by the end of the novel Tom will take Casy's place as a breacher of exploitative fences:

> "They's gonna come a thing that's gonna change the whole country."
> Tom said, "I'm still layin' my dogs down one at a time."
> "Yeah, but when a fence comes up at ya, ya gonna climb that fence."
> "I climb fences when I got fences to climb," said Tom.
> Casy sighed. "It's the bes' way. I gotta agree. But they's different kinda fences. They's folks like me that climbs fences that ain't even strang up yet—an' can't he'p it."[6]

Casy already anticipates legal and extralegal barriers the migrants will encounter, actual and metaphorical fences they will have to climb if they want to escape the brutal greed of exploitative landowners and the inequitable nature of the economic system as a whole. Elia Kazan's 1952 film *Viva Zapata*, for which Steinbeck wrote the screenplay, again employs barbed wire in a conflict between peasants and the wealthy landowners who have appropriated the peasants' land. The peasants petition President Porfirio Díaz for help. He says they must have proof. The land has been newly fenced with barbed wire, and when the peasants cut the fence to get to the stones that mark the land, they are swept by machine-gun fire.[7]

Just a few pages after the exchange between Jim Casy and Tom Joad, Tom and his brother Al are talking about Tom's time in prison, and Tom begins to question the idea of prisons and the idea of locking up human beings: "Somepin screwy about it, somepin screwy about the whole idea a lockin' people up." They approach a wrecking yard enclosed with barbed wire: "The truck drove to the service-station belt, and there on the right-hand side of the road was a wrecking yard—an acre lot surrounded by a high barbed-wire fence. . . . In the grassy lot the wrecks lay . . . a mass of derelicts." The fences that failed to protect their own property here enforce the rights of a boss who has left a much-abused one-eyed man in charge. As opposed to the resourceful Joads, however ill fated, the man in the wrecking yard is simply pitiful. Tom advises him to put a patch over his empty eye socket and to leave: "'Whyn't you roll on? Got no guards to keep ya here.' 'Yeah, that's easy to say. Ain't so easy to get a job—not for a one-eye' man.'"[8] Subjugated by the fences that have invaded his mind, there will be no rolling on for this man.

The moral of the barbed wire story always depends on who is stretching the fence. While fences marked but failed to protect the Joads' land in Oklahoma, fences protect working families banded together in the California government camp the Joads find after being chased out of an unprotected Hooverville: "A high wire fence faced the road, and a wide-gated driveway turned in." "Some nights the boys patrol the fences, 'specially dance nights,'" the watchman explains, a defensive maneuver necessary because landowners resent and fear the migrants who have organized the camp for their own well-being. When the Joads end up behind the fences of a landowner's camp, the fence and attendant armed men are there to control both the migrant laborers inside and the angry picketers outside. Tom decides to investigate but is stopped by a guard with a gun and a flashlight who asks where he is going: "Well, I thought I'd take a walk. Any law against it? . . . Can't I even get out of here?"[9] There is in fact no law against it, but economic conditions and capitalist exploitation of the migrants' labor make these barbed wire fences possible and necessary. Tom circles through a field:

> At last he came to the wire fence, five strands of taut barbed wire. Beside the fence he lay on his back, moved his head under the lowest strand, held the wire up with his hands and slid himself under, pushing against the ground with his feet.
>
> He was about to get up when a group of men walked by on the edge of the highway. Tom waited until they were far ahead before he stood up and followed them. . . . Tom climbed a fence and moved down into the ravine through brush and dwarf willows; and in the bottom, beside a tiny stream, he found a trail.[10]

With other strikers, Jim Casy is camped beside the stream, inhabiting a liminal space beyond the two fences. He and Tom have crossed those fences to enter a dangerous but potentially transformative place. Beyond the fences they are free. They are also unprotected. Moments later they are struck down by men carrying ax handles. Casy is killed. Tom fights back, possibly killing one of the assailants, and then escapes, recrossing the boundaries into the camp: "Ahead he saw the bushes that bounded the field, bushes along the edges of an irrigation ditch. He slipped through the fence, edged in among vines and blackberry bushes. And then he lay still, panting hoarsely."[11]

The Joads spend the rest of the novel outside organized camps, Tom in hiding, the rest of the family in a boxcar until they are finally flooded out. Fleeing the rising water, they cross a final fence to get to the barn where the novel leaves them: "'Le's go through the fence here. It's shorter. Come on, now! Bear on Rosasharn.' They half dragged the girl across the ditch, helped her through the fence." Beyond this fence lies the mythical opening into shared spirit that Jim Casy preached and that Tom embraced: "Well, maybe like Casy says, a fella ain't got a soul of his own, but on'y a piece of a big one." Rose of Sharon, having just given birth to a dead child, breaches a psychological fence and bares her breast to nurse a starving man: "She looked up and across the barn, and her lips came together and smiled mysteriously."[12]

As the Joads and Jim Casy "climb" the various fences (tangible or otherwise) meant to exploit them, they risk their lives to keep moving. They leave the restraints of prison and jail and Hooverville and the landowner's camp to arrive at a powerful solidarity calculated to change the conditions that have made them desperate. They also climb internal fences. And here we return to Jim Casy's comment about barbed wire:

> Why is it that when a fella ought to be just about mule-ass proof against sin, an' all full up of Jesus, why is it that's the time a fella gets fingerin' his pants buttons? . . . I says, "Maybe it ain't a sin. Maybe it's just the way folks is. Maybe we been whippin' the hell out of ourselves for nothin'." An' I thought how some sisters took to beatin' theirselves with a three-foot shag of bobwire. An' I thought how maybe they liked to hurt themselves, an' maybe I liked to hurt myself.[13]

These three-foot shags represent a "different kinda fence," a psychological fence Jim Casy is eager to transcend. Casy decides to quit preaching. He does not know anyone named Jesus, he says, but he does know plenty of real people who need his help. The ecstatic self-mortification and emotional excesses linked to the "Holy Sperit" occur repeatedly in the novel and are tied to the double reasoning Casy lays out here. The people beating themselves do so because they feel they themselves are the sinful causes of their miserable conditions and thus must literally beat the hell out of themselves. Or, perhaps, they just like to hurt themselves. Examples of self-harm in response to perceived guilt abound in the novel.

Ma Joad repeatedly rebuffs the howlers and jumpers who choose ecstasies of the "Sperit" over human solidarity. When Granma is dying in a tent on the California border, a "Jehovite" pushes her way into the tent and announces that she will "hol' a meetin'—a prayer an' grace" for the "soul here ready to join her Jesus. Praise God!" Ma Joad says coldly, "We ain't gonna have no meetin' in this here tent," and the woman returns to her own tent, where she and her fellows raise exhortations that become feral howlings, screams, slapping, and thudding. Ma Joad questions her own decision to turn the Jehovites away but then tells Rose of Sharon, "Maybe it's him [Jim Casy] made me tell them people they couldn't come here."[14] The howlers may goad themselves to ecstasy, but they do nothing to change the conditions that have forced the Joads and others from their homes.[15] In the government camp, Ma Joad confronts another religious fanatic who has frightened Rose of Sharon because she "went out to a meetin' in Weedpatch las' night. Know what the preacher says? He says 'They's wicketness in that camp.' He says, 'The poor is tryin' to be rich.' He says, 'They's dancin' an' huggin' when they should be wailin' an' moanin' in sin.'" Ma Joad threatens her with a stick, and she begins to howl and drool and shudder and falls to the ground in a fit. A watching man attributes it to "the sperit. She got the sperit." Ma Joad will have none of that: "If she comes back, I might hit her." Uncle John is also obsessed with sin: "Nobody don' know my sins, nobody but Jesus. He knows."[16] His own version of penance involves alcohol. Two bottles of whiskey bring him down as surely as the woman felled by the "sperit." In this novel, all attempts to appease an angry God through self-punishment are self-defeating.

Aroused by sin and sorrow, facing insufferable conditions, people slap and howl and flagellate themselves with barbed wire. Rather than whip themselves with barbed wire, rather than numb themselves with spirits to the conditions under which they suffer, Jim Casy argues that because the enemy is an unjust and oppressive economic system enforced by people who profit from it, the battle can be won only through human solidarity and organized labor.

Karl Marx, of course, referred to religion as the inverted consciousness of the world and argued in terms that resonate in *The Grapes of Wrath*: "Religion's suffering is, at one and the same time, the *expression* of real suffering and the *protest* against real suffering. Religion is the sigh of the oppressed creature, the heart of a heartless world, just as it is the spirit of spiritless conditions. It is the *opium* of the people."[17] Georges Bataille describes the process similarly: "Religion thus does not at all try to do away with what others consider the scourge of man. On the contrary, in its immediate form, it wallows in a revolting impurity that is indispensable to its ecstatic torment . . . in a mental agonistic orgy practiced at the expense of the real struggle."[18] This is precisely Jim Casy's argument. Self-flagellation with barbed wire is a "mental agonistic orgy practiced at the expense of the real struggle."

Flannery O'Connor begs to differ.

Wise Blood

If it's a symbol, to hell with it.

—Flannery O'Connor, in conversation with Mary McCarthy

"Mrs. Tate had to tell me once that there was no such thing as bob-wire. It is barbed wire. Isn't that silly? My mother says, 'You talk just like a nigger.'"[19] When she says "bob-wire," Flannery O'Connor also talks like the characters in *The Grapes of Wrath*. There is surprising congruence between Steinbeck's antireligious novel and Flannery O'Connor's grotesquely mystical *Wise Blood* (1952), set in the fictional town of Taulkinham, Tennessee. The most notable parallel is the mortification of the flesh by barbed wire featured in both works. *The Grapes of Wrath* begins with Jim Casy's renunciation of the ecstatic Christianity that leads women to whip themselves with barbed wire, and *Wise Blood* ends with Hazel Motes's vicious self-cleansing by wrapping his chest with barbed wire. A striking set of parallels lends credence to our reading of O'Connor's barbed wire as deliberately antithetical to Steinbeck's.[20] As the table demonstrates, Motes enters *Wise Blood* in step-for-step congruence with Tom Joad in *The Grapes of Wrath*, additional parallels accrue, and the self-harm with barbed wire provides an ultimate contrasting commonality.

O'Connor claimed not to have read Steinbeck. In a letter to "A." on August 21, 1955, she wrote disparagingly of a Nelson Algren book she was reading and used Steinbeck as an example: "A writer . . . cannot sentimentalize the poor

Parallels between *The Grapes of Wrath* and *Wise Blood*

The Grapes of Wrath	*Wise Blood*
Tom Joad returns after four years in prison.	Hazel Motes returns after four years in the army.
He wears prison-issued new clothes: a "stiff" gray cap and a suit of cheap gray hardcloth.	He buys a cheap, new suit and a "stiff" hat.
He catches a ride in a truck.	He catches a ride in a cotton-seed truck.
He finds the family house pushed off its foundations, the fences gone, and the cotton growing in the dooryard.	He returns to a shell of a family house, the fence around it partly fallen, and weeds grown through the porch floor.
Jim Casy comments on Jesus: "No, I don't know nobody name' Jesus."	The train porter opines that "Jesus been a long time gone."
Casy wears sneakers, "scared there's wire or glass under the dust."	Motes put glass and stones in his shoes to hurt himself.
A newly acquired but hard-used truck carries the Joads to a new life.	Motes buys a dilapidated old car and attempts to flee to a new town.
Tom Joad and Casy are fence climbers.	Motes crosses a barbed wire fence.
Casy renounces ecstatic Christianity and self-flagellation with barbed wire.	Motes practices Christian self-cleansing by wrapping his chest with barbed wire.

and get away with it. . . . I have never read JT Farrel or Steinbeck or any of the people who deal with the afflicted (econimically [*sic*] afflicted that is)."[21] Her dislike of Steinbeck is also apparent in a letter of November 3, 1962, in which O'Connor asked "A.": "Can you fancy Steinbeck getting the Nobel prize. John O'Hara will be getting it next."[22] Her detailed protestation of ignorance suggests that O'Connor has indeed read Steinbeck, and her statement of dislike indicates that she knows the work well enough to denounce it. A devout Catholic, O'Connor had good reason to rewrite a novel in which a preacher turns from Jesus to the people. The fact that Jim Casy is presented as what Theodore Ziolkowski calls a "comrade Jesus" (his initials, his identification with the Jesus who went into the wilderness to find his calling, his journey west with the twelve Joads, his sacrifice of himself for Tom Joad and for workers in general) must have been especially irksome to the deeply religious O'Connor.[23] In our reading, O'Connor wrote her novel as an antidote to Steinbeck's godless work and set up the parallels we have noted to make that explicit.

Hazel Motes's obsession with sin is related to Steinbeck's self-hating, barbed wire–wielding women. At a carnival as a boy, Motes witnessed his father leering at a lively naked woman in a silk-lined coffin. He tried to atone for having seen that by putting rocks in his shoes. In the army, he settled on a different approach, reasoning that if he did not sin, a threatening Jesus could not get him. Motes's belief in divine violence comes in part from his grandfather, a preacher, a "waspish old man . . . with Jesus hidden in his head like a stinger." Out of the army, he takes a different tack. If there is no Jesus, he decides, then there is no sin: "You are clean without Jesus." The blasphemies Motes utters, his fornication with a whore and with a young girl, and his murder of the false prophet with his car are sinless acts if there is no Jesus to call them sin. Finally, however, his car gone and his deeds done, he succumbs to "the wild ragged figure" that has flitted the whole time in the back of his mind.[24] He pays for his sins by blinding himself, wrapping his torso with barbed wire, and lining his shoes with gravel and broken glass.

One morning Motes's landlady finds him breathing heavily:

> The old shirt he wore to sleep in was open down the front and showed three strands of barbed wire, wrapped around his chest. She retreated backwards to the door and then she dropped the tray. "Mr. Motes," she said in a thick voice, "what do you do these things for? It's not natural."
>
> He pulled himself up.
>
> "What's that wire around you for? It's not natural," she repeated.
>
> After a second he began to button the shirt. "It's natural," he said.
>
> "Well, it's not normal. It's like one of them gory stories, it's something that people have quit doing—like boiling in oil or being a saint or walling up cats," she said. "There's no reason for it. People have quit doing it."
>
> "They haven't quit doing it as long as I'm doing it," he said.
>
> "People have quit doing it," she repeated. "What do you do it for?"
>
> "I'm not clean," he said.[25]

This penance for uncleanliness is extreme, but it is congruent with a history of religious violence. And barbed wire (the thorny fence with symbolic connections to Christ's crown of thorns) is a perfect instrument for self-harm.

Whether Hazel Motes has become a saintly believer in harsh penance or whether he has proven that he is crazy is a disputed topic among critics.[26] But by the internal logic of the novel, he has moved from a childhood fear of a sharp-stingered Jesus to an attempt to establish a church without Christ (and thus without sin and the attendant penance) and finally to an extreme sense for his own sinfulness that requires extraordinary punishment. The landlady sees the acts as unnatural, attributing them to gory stories from the past. Jim Casy similarly relegated talk of Jesus and penance to stories from the past: "No, I don't know nobody name' Jesus. I know a bunch of stories, but I only love people."[27] Motes, believing that Christ is not a character in a story but a real being who demands real penance for his sins, interprets the brutal acts of penance as absolutely natural.

Like Motes, O'Connor herself believed in the reality of the Christian stories. Brad Gooch describes a conversation at the New York apartment of novelist Mary McCarthy during which McCarthy spoke of the host taken at communion as a symbol of Christ's body and O'Connor replied: "If it's a symbol, to hell with it." Gooch points out that O'Connor's friend, the Catholic poet Robert Lowell, mortified his flesh by entering a cold bath on hands and knees on the day of O'Connor's patron saint St. Therese of Lisieux. Some details of Motes's gory penance, according to Gooch, came from O'Connor's inscribed copy of Robert FitzGerald's *Oedipus Rex* and from her reading of the works of three saints—Catherine of Sienna, Catherine of Genoa, and Theresa of Avila.[28] Catherine of Sienna, in her *imitatio Christi*, famously scourged herself and died at the age of thirty-three after denying herself food and water. Catherine of Genoa wore haircloth, ate nothing for the season of Lent, and slept on sharply pointed objects. And Theresa of Avila, joining Christ in his passion, assiduously mortified her flesh. When Motes follows suit, he proves that the harsh asceticism of the past is alive and well in the twentieth century, contrary to the landlady's supposition.

For readers alerted to the motif of barbed wire in the novel, there remains the scene in which Motes finds Sabbath Hawks in the backseat of his car. He had earlier decided to seduce her, and she now has intentions herself. While they drive, she announces that she is a bastard and thus "can't enter the kingdom of heaven." Motes is dumbfounded: "You couldn't be a bastard. . . . Your daddy blinded himself." By Motes's flawed but penitential reckoning, blinding oneself is full recognition of and payment for sin. Therefore, a man who could perform such a righteous act could never be the father of a bastard. They get out of the car, and Motes has a question: "'Was he a very evil-seeming man before he came to believe,' he asked, 'or just part way evil-seeming?' 'All the way evil,' she said, going under the barbed wire fence on the side of the road." Inside the fence, Motes thinks about the bastard question while Sabbath does her best to seduce him. When she looks at him from very close range and says, "I see you," he jumps up and returns to the car, and she "came running up to the fence. She got down on the ground and rolled under the barbed wire and then stood at the window of the car, looking in at him."[29] It is a curious scene, but our focus on barbed wire is helpful. Enclosed by barbed wire fences, potentially "seen" in his wickedness, Motes flees the enclosure and returns to his wreck-of-a-car and to his church without Christ. Refusing to be seen, he remains, for some time, a fugitive outside the barbed wire bonds of Christian dogma. He recants, finally, and mortifies his flesh with the very barbed wire from which he here escapes.

If blinding, barbed wire, sharp glass, and rocks are the means to Hazel's purification—a self-crucifixion of sorts, an *imitatio Christi*—then the earlier appearance of the instruments of penance in the novel might be seen as foreshadowing the final horrors, stations of the cross, as it were: Young Motes puts rocks in his shoes to punish himself; he encounters the man who supposedly blinded himself; he crosses and recrosses the barbed wire fence with Sabbath; and finally he raises himself on the "cross" of quicklime, barbed wire, and rocks and glass in his shoes. There is an echo here, with the wire and glass, of Jim Casy's admission that he puts on his canvas sneakers because "I ain't got your confidence. . . . I'm always scared there's wire or glass under the dust."[30] While Casy puts on his shoes to enable his work for the people who embody holiness for him, Motes puts on glass-filled shoes to do penance.

Jim Casy moves away from God and the "Holy Sperit": "Why do we got to hang it on God or Jesus?"[31] Motes returns to Christ with a vengeance. Jim Casy says the sins are not real. They come from stories that need to be superseded; self-flagellation is unnatural; human solidarity against economic exploitation is natural; the "grapes of wrath" should turn us away from Jesus, who "died to make men holy," and toward sacrifices that will "make men free." Hazel Motes decides that the sins are real, the stories about Jesus and his penitent followers are real, and self-flagellation is "the fateful lightning of his terrible quick sword."[32] Whether it is a "three-foot shag of bobwire" or "three strands of barbed wire, wrapped around his chest," the "thorny fence" lends itself readily to metaphors related to Christ's crown of thorns and then, for better or worse, to reification of the metaphor on the bodies of ecstatic "saints."

"Contagion"

Words must serve in the stead of a fence.

—Brian Evenson, "Contagion"

Brian Evenson's story "Contagion" continues the masochistic religious theme Steinbeck and O'Connor have explored and does so in a more conventional setting, more conventional, at least, for stories involving barbed wire. Set somewhere in what we suppose is the American Southwest, the story begins with a paragraph about the initial task:

> They were to travel due South, checking fenceline for $2/day to territory's extreme, and then to cross over and observe conditions beyond. They rode by horse, seeing only perfect and secure fence to either side of the road until, near the border, the landscape grew ribbed and strange. It was not merely stony but gnarled all through like a brain.[33]

Cowboys riding fence for two dollars a day. We know this story well. There will be a range war. Fence cutting. A choice between fenced civilization and wilderness. But then the second sentence transports a reader across a border and the "perfect and secure fence" gives way to a strange landscape "like a brain." Like much of Evenson's work, "Contagion" is about what goes on in a brain, about phenomenology and epistemology. In their search for certain knowledge, for the control that absolute knowledge might bring, his characters either embrace webs of lies or live outside the seductive constraints of civilizing language.

The phenomenon to be apprehended in this story is contagion. The most terrible contagion, however, as the story's epigraph from Artaud informs us, is the contagion that does not reveal its traits. The contagion of this story is indeed beyond understanding, so the fence riders consider the things they can apprehend—the barbed wire fences they have been paid to check.

The three company men bear the names of protagonists in barbed wire history. Early in the story, Hunt joins his horse as a victim of the inexplicable contagion and by telegraph the company orders Haish and Grenniger to "*pursue contagion along fenceline stop else all monies forfeit stop.*"[34] The task thus shifts from checking the actual fence to pursuing the contagion along the fence line. Straddling the line between the thing itself (the barbed wire fence) and the thing as metaphor (the meaningful fence), Haish, even as he continues to make a simple record of what he sees, begins to contemplate the meaning of the fence:

> The fence changed and Haish promptly recorded it in his book. *Double-strand undulate wire, coldweather, armatured with Glidden's coil. . . . What are we searching for exactly?* he wrote. *What might we have to learn from wire? Is it better here, across the border, among new varieties, hand-patched? What is the connection between wire and contagion, if any? When will I die?*[35]

The obvious answer is that there is no connection other than circumstance. There is the fence. And there is contagion. The contagion does not reveal its traits. Barbed wire has only material traits, but Haish is driven to know and begins to construct meaning out of what he has at hand: barbed wire. Haish knows he is "entering a speculative space he should shy from," but he continues to try to understand the fence, "pursuing the paths of his own brain."[36]

The two riders continue their company-imposed task. They come across an animal impaled on the fence, a "shoat, wedged between the bottom wire's spurred wheel and the ground, three bloody grooves carved into its back. It was half-standing, the spurs of the wheel deeply embedded." Later they see something on the fence that at a distance looks like a scarecrow: "He was dead, fence-strung, the barbs hooked by design into his shirt and belt and keeping him afloat on the wire." Based on a pouch at his waist full of wire and a barbing tool, they speculate that he was a company man like them. Perhaps they too are at risk. The fence is a useful, if vicious, tool for keeping animals in or out; it is also potentially lethal if the fence is contested. Once again, readers of Westerns are on semi-familiar range here, until, that is, the riders enter a small town where grim-faced men ask them if they "follow the fence?"[37] When they say they do, an admission they mean literally, the town's residents understand them figuratively and shout praises. The fence riders, it turns out, have come across a sect of fence worshippers whose leader is named Glidden.

Haish and Grenniger are loath to show their company notebook to Glidden, but he gradually overcomes their opposition and reads and interprets the notes Haish has made. Grenniger asks Glidden about the contagion, and Glidden replies that "what you think of as contagion is merely an extension of wire, for not all wire is physical. This contagion is no more than animals bound in a wire too fine for their vision to perceive."[38] The thing, the wire, becomes metaphor to explain what cannot be explained. Most abstractions have their beginnings as something physical, things that can be used as metaphor to mean something else. Catholic mystics, for example, attempting to represent their union with the divine, almost invariably turned to the imagery of sex. This is precisely the process of truth making Nietzsche wrote about: "What is truth? a mobile army of metaphors, metonymies, anthropomorphisms . . . only insofar as man forgets himself as . . . an artistically creative subject does he live with some calm, security, and consistency."[39] As a tool in the real world, barbed wire controls, separates, and imposes order. In response to the incomprehensible and frightening contagion, the town's panicked populace transforms the wire's function until the fact of the wire becomes the truth of the new religion: "You shall know the fence and the fence shall make you free."[40] Move from thing to metaphor, forget that it is a metaphor and call it the truth, and the truth will free you from the inexplicable contagion.

Grenniger disappears. Haish is drugged and kept in a room with the notebook and a pencil and told to write about the fence: "*You are but an instrument*, Glidden sometimes said, and, of the notebook, *this work shall be one of the chief posts from which the wire is strung. . . . Words must serve in the stead of a fence.*" To inspire him, Glidden takes Haish to a warehouse that houses what amounts to a religious museum of barbed wire: "The walls were run with strands of barbed wire of all types and all varieties, an elaborate armature for the room. To see so much wire together rather than sparsed out along miles of fence made Haish's head purl." There follows a remarkable full-page account of the kinds of barbed wire making up the armature ("splint barb, nail barb, knife-edged t-bar . . . caduceus, spiral, prong, kink"), a list that ends as a subtle comment on the inevitable result of encounters with both real and metaphorical barbed wire: "snake tongue, knee grip, offset, seated, webbed, caged, trapped."[41]

Caged and trapped by the worshippers of barbed wire, Haish begins to circle the walls of his room with a running line of text. He writes an imaginative account of Grenniger still following the fence, encountering fence cutters, and finally coming to the end of the fence where he "beheld a vast and fenceless and barren plain, unbordered and running on forever. *There was no source to the contagion, he realized. . . . It was simple and endless*."[42] Simple and endless. Not to be understood or comprehended by a set of beliefs based on barbed wire metaphors. Not to be misapprehended and misrepresented through the language of religion. As surely as the actual barbed wire strands they resemble visually, the words and sentences running along his walls control Haish's body and drug-hazed mind—"for not all wire is physical." A reader does not initially realize that Haish is the narrator of this account of Grenniger's action, alerted to that fact only when Haish stumbles verbally back into the story: "*There was no source to the contagion, he realized,* Haish wrote. *It was simple and endless.*"[43] The break from omniscient narration into a story narrated by one of the characters reminds a reader that even a text that warns of sentences and stories that bind and control is itself an enclosure. Nietzsche's "prison house of language" cannot be transcended by language.[44] At most, a text can point to its own threatening character.

Finally, angry at a supposed heresy, Glidden takes Haish (who is suffering the first bloody signs of contagion) back to the warehouse, strings him up by his hands in a kind of crucifixion, and wraps his chest with barbed wire (in the tradition of Steinbeck's women and of Hazel Motes, and of the other members of the cult whose scarred torsos Haish sees crowded around him). Glidden explains the theology behind his actions: "There is the physical wire and the spiritual wire. . . . The body knows to heal from the one but it must be coaxed to recognize the other. We must fight wire with wire." Haish sinks into a meditative state and begins to "sense the structure of the contagion itself, the hidden system of post and wire." He contemplates the physical wire and the spiritual wire, the body that is the post and the brain that is the wire: "What was left of him waited to see how long it would take for post to give way, fence to collapse."[45]

And with that Evenson joins Steinbeck in using barbed wire as a metaphor for the ideological and religious structures we create through language, social constructs as barbed and controlling as is the thorny fence for animals and humans alike. Are we company men and women controlled by the barbed wire? Or are we fence cutters? Are we so afraid of contagion that we are willing to inflict the pains of barb-enforced quarantine on others and on ourselves? Or are we heretics of the fence, clear about the fact that any fence that protects us simultaneously imprisons us? Evenson's story depicts a coercive sect that captivates its members, transforms barbed wire into metaphors that imprison their minds, and then binds them with real barbed wire and with controlling metaphors as a cure for the inscrutable contagion. The history and artifacts of barbed wire can be as useful in the construction of meaning as the history and artifacts of a crucified Jew.

"A History of Barbed Wire"

My tattoo, first tattoo. Black barbed wire.

—Jeff Mann, "A History of Barbed Wire"

"A moral here we would inspire / To dudelets all, both son and sire: / In spreading on your agony dire, / Beware the *peel*—and Glidden wire— / They're death on dudes." The "Dudelet," published in the 1885 *Glidden Barb-Fence Journal*, brings two effeminate characters to abrupt contact with a spool of Glidden barbed wire that proves to be "death on dudes."[46] Barbed wire is a masculine tool, the poem declares, not for sissies. The unnamed author of "The Dudelet," a product of the time and a servant of the contemporary advertising medium, could not anticipate Annie Proulx's "Brokeback" cowboys, much less the dominant/submissive masculinity of Jeff Mann's story "A History of Barbed Wire." Steinbeck's Jim Casy understood the phenomenon: "An' I thought how maybe they liked to hurt themselves, an' maybe I liked to hurt myself." Hazel Motes and cult-leader Glidden may have understood as well.[47] None of the three characters, however, would have comprehended how a positive identity might be established through such pain.

Mann's narrator has a new tattoo: "Black barbed wire wrapped tight around my right biceps and triceps." The tattoo represents "a desire which took decades to distill," and the story itself is a history of that distillation.[48] The narrator's first memory in this regard is from his boyhood in West Virginia:

> I'm helping my father run barbed wire along a new pasture fence, from locust post to locust post, then about a corner oak. We tug the thorny wire carefully off its roll, pull it tight, and staple it to the wood. Always there's the knowledge, the dread, the slightest chance that a tautly stretched length of wire might snap. I remember stories: the farmer in an adjoining hamlet, the sudden accident, the way the wire, too tense, turned on him just as a rattler abruptly betrays the snake handler's confident piety.[49]

The memory is fraught with danger, especially when his father orders him to pull the wire tighter: "I start nervously, and one of the barbs pierces my thumb." It is also redolent with sex. He imagines the neighboring farmer, naked, injured by the snapped wire, wrapped and bleeding: "The barbed wire's wrapped about his torso, dark hair on his chest is rilling with rain and scarlet."[50]

In high school the narrator reads about gay love and finds himself attracted to the local mountain men but knows they would respond to him with contempt and violence. He wonders what manhood means in light of his desire and decides he wants "to become the sort of man I desire, I want somehow to absorb the apparently effortless masculinity of country boys. Boots, beards, chest hair. Denim jackets, pickup trucks. Tattoos."[51]

He gradually finds his way into dominant and then submissive sex. And now, as he tells his story, he is in St. Mel's Church in Longford, Ireland, near the graves of ancestors, thinking about how he is aging: "Before me, Christ opens his robe to reveal his heart, which is tightly wrapped in thorns." The sight takes him back to the corner oak around which he and his father stretched their fence: "Barbed wire again, its intermittent fangs my father and I wrapped around the tree's expanding girth." He imagines the wire gradually swallowed in the tree, "a sharp suffering disappearing beneath the skin." He also imagines his last lover in the position of Christ on the cross: "To the bed, to this crucifix of sheets, I tie him tight—the way we both like it—then hurt him gently."[52] He laments how he has become domesticated and imagines "some reckless, irrevocable gesture, some proof that a little youth, a little edge is left." And thus the tattoo: "A small proud pain, almost a pleasure: nails, honeybees, wild rose thorns. . . . Now barbed wire's tied about my arm. Appetite's inscribed, flesh is emblematic."[53] He sees the tattoo as indelible desire, as permanent passion: "Thorns wrapped around the heart of the Sacrificed God. A naked man bound tight . . . learning the pedagogy of suffering."[54]

And with this we have come full circle, back to the ecstatic women whipping themselves with barbed wire because they like it. With his crown of thorns and his bleeding sacred heart, Jesus finds a modern counterpart in the tattoo of the thorny wire. As is the case for other rural stretchers of barbed wire fences, barbed wire is the metaphor right at hand for Mann's narrator, rich with tension and blood, a metaphor that can be reified in bed or in the ink of a tattoo. Barbed wire is the devil's hatband, the devil's rope, the devil's wire, the savior's crown of thorns.

CHAPTER SEVEN

"I Helped Him Build His Own Fences"

Cutting the Wire, Cutting the Lies

He had only seen and heard the world as it always was: no boundaries, only transitions through all distances.

—Leslie Marmon Silko, *Ceremony*

This is my home
this thin edge of
barbwire.

—Gloria Anzaldúa, *Borderlands/La Frontera*

Early in D'Arcy McNickle's novel *Wind from an Enemy Sky*, Henry Jim visits his brother Bull after a long estrangement. In a speech given in Bull's tepee, he remembers the time when the Little Elk people could travel freely in any direction. Now, he says, things are different: "If we traveled out today as we did in those times, we'd be stopped by a fence, or a railroad, or a highway, all fenced." He apologizes for his hand in the changes: "I went to the white man, I spoke for him. I opened the gate and let him come in. I helped him build his own fences." Bull finds it difficult to reply to his brother: "How could he push aside thirty years as if they had never troubled him or the people who stayed with him in their foothill camps and went hungry with him because he would not lead them down to the flats where they could have had government wagons and teams, flour and coffee, fence posts?"[1] Henry Jim has all of those things because he lives on the flats with fields enclosed by government barbed wire. While the split between the brothers is symbolized by the contrast between the fenced flatland and the unfenced foothills, the deeper enmity between them is the result of Henry Jim having given their strongest medicine bundle to a white man who placed it in a museum.

In this chapter we explore the barbed wire motif in works by Native Americans and by a woman who identifies as mestiza ("I will have my voice: Indian,

Spanish, white").[2] Barbed wire is a multivalent signifier of Native identity, identity both imposed and asserted. But first we present some historical context that illustrates how barbed wire was used to take over and control land inhabited by Native Americans. Ironically, once the Indians were safely controlled on reservations, barbed wire became a government commodity that Indian agents could award or withhold, depending on the behavior of their charges.

Historical Barbed Wire from Native Perspectives

Henry and Frances McCallum tell a story drawn from H. T. Burton's "A History of the JA Ranch" and J. Evetts Haley's *Charles Goodnight* in which Pueblo Indians on a trading mission from New Mexico to Oklahoma are having trouble finding their way back to Taos. They tell Goodnight, who has intervened on their behalf in a potentially dangerous standoff with members of a new Christian Temperance Colony in the Texas Panhandle, that they do not know how to get back to Taos. Goodnight says that is impossible; they know the country from a life of experience. The Pueblo answer is "Alambre! Alambre! Alambre! Todas partes!—wire! wire! wire! everywhere!"[3]

Native Americans lost lands and migratory cultures to the stringers of barbed wire who had been promised by advertisements that their savage fences would protect them from the Indians. Edward Curtis, most famously, photographed Natives in an attempt to document traditional life before it disappeared, and Joseph Dixon, funded by the Wanamaker family, also took hundreds of photographs of Native Americans to capture traces of what he saw as a "vanishing race." The task was difficult, in part, because barbed wire now dominated the landscape. Shining strands of barbed wire are evident in Dixon's photograph *Native Americans on Horses*, as is the attempt to scratch the fence posts off the negative (figs. 35 and 36).[4] It is difficult to feel good lamenting the passing of the romanticized Indians when confronted with one instrument of their eventual conquest. Dixon attempted to solve the problem with his crude fictionalization of the photo.[5]

A striking example of how barbed wire was used as a tool of subjugation can be found in US Indian agent John P. McGlinn's 1890 "Report of Neah Bay Agency." After mentioning an influenza epidemic, reporting on a severe winter during which cattle and horses died in large numbers, and detailing facts about the reservation farm, McGlinn turns to his experience with the Quillayute Indians. He notes that in 1889 President Grover Cleveland had set aside

Figure 35. *Native Americans on Horses*. Photograph by Joseph Dixon. Courtesy the Denver Public Library, Western History Collection, Z-3224.

Figure 36. Close-up, Dixon's *Native Americans on Horses*.

more than eight hundred acres as a reservation for the Quillayutes, "provided that this withdrawal shall not affect any existing valid rights of any party."[6] Unfortunately, McGlinn writes, the eight hundred acres had already

> been taken up previously by whites under the homestead and preemption laws. Not an acre that is worth anything to them is left. Their village, their homes, and what has been the homes of their fathers for generations, as the immense shell mounds prove, has been homesteaded by a white man, who has erected his dwelling-house in the center of this village.
>
> Shortly after the Quillayute Indians left their village last September, on their annual pilgrimage to the hop-fields of the Puyallup Valley,

> twenty-six of their houses were destroyed by fire, with all they contained, consisting of whale and fur-sealing outfits, canoes, oil, etc. After the fire Mr. Pullen, the settler, sowed grass-seed on the site of the burned homes, inclosed it with a barbed-wire fence, and not satisfied with doing this, fenced them off from every other available location by five strands of barbed wire. . . . Being fenced off from the hill, they were compelled to erect their new houses on the beach, where they are very much exposed to the fury of the ocean.[7]

Sympathetic to the Quillayutes' plight, McGlinn laments that barbed wire—coupled, he says, with homestead and preemption laws—has been used to exclude the Natives from their ancestral homelands. The five-strand fence stands as a marker of property, enforcing legal rights of conquest.

At this point in the report, McGlinn turns to a problem of a different sort: Quillayute culture. "All these coast Indians," he writes, "are as superstitious as the natives of Central Africa. . . . The adult Indian knows comparatively nothing regarding religion or morality."[8] As a result, he has rescinded previous practice that sent children home from the boarding school for weekends and is keeping them under close supervision so they will not be influenced by the "heathenish and barbarous practices" of their elders.[9] As we saw in Brian Evenson's "Contagion," "there is the physical wire and the spiritual wire."[10] Barbed wire fences appropriate Native land; the boarding school appropriates Native culture. The latter is as much an act of conquest as the former, as Gerald Vizenor points out: "The causal narratives of missionaries and ethnologists are terminal simulations of dominance, not survivance."[11] We return to the idea of survivance shortly.

Native Americans began to use the fence themselves even as the fence was used against them, as a 1908 Senate hearing on the allotment of Indian lands demonstrates. The hearing was precipitated in part by the revelations and provocations of Helen Pierce Grey, an activist journalist who had originally gone to the Crow Reservation in Montana for a more general story but had uncovered, she claimed, corruption and theft of Indian land. Grey testifies about attempts by the Lincoln Townsite Company to acquire allotment lands by various forms of coercion. She spoke with President Theodore Roosevelt about it, she says, and the lands were then deeded to the US government. Regardless of that action, sheep were grazed on the reservation in such numbers that the Crows were unable to grow crops and were forced to move.[12]

Samuel G. Reynolds, the US Indian agent whose practices are under investigation, is asked patronizing questions about the Indians' ability to farm, about whether any of them are "like the old-time blanket Indians, whether they wear blankets or citizens' clothing."[13] Testimony reveals that Reynolds had Helen Grey arrested three times for disturbing the peace and for being on the reservation without permission. George W. Hogan, a Crow Indian, testifies that he too had been put in jail by Reynolds but released on the fifth day after he threatened he would demand an investigation of the agent:

> He was set at liberty by Agent Reynolds, who told this affiant that if an inspecting official came to the reservation that this affiant must attend to his own business, and that he, the said Agent Reynolds, would furnish this affiant with all the barbed wire he wanted and aid him in every manner. This affiant understood this to mean a recompense for the unjust imprisonment and to influence this affiant to make no complaint when any inspecting official visited the said Crow Reservation.[14]

The suitability of barbed wire as a bribe in this case is highlighted in a letter to the Department of the Interior, May 10, 1905, sent by "Plenty Coos [*sic*]," chief of the Crow Indians and cited in the hearing. Plenty Coups makes several requests, the fourth of which regards barbed wire:

> My young men are handicapped in their efforts to farm their lands. Although we have been supplied from time to time with barbed wire to fence our lands the supply is insufficient for the number of people on the lands where they do their farming, and the young men who become of age are being allotted and their lands are not fenced. They are willing to work their lands, but not having their lands fenced they are very badly handicapped and discouraged in their efforts to raise crops. Therefore, I request that some wire be purchased—three carloads—and sent to our agent for him to distribute to the ones that need it. . . . The $40,000 in the agreement of 1904 is for the purpose of fencing the entire Crow Reservation, and, of course, is not available for individual allotments.[15]

Z. Lewis Dalby, sent to inspect the agency after Grey's accusations, notes:

> In her charges Mrs. Grey alleges that the agent shows favoritism in the distribution of issuable material, particularly barbed wire, charging that he gives these things out as rewards for "packing news" in connection with the "spy system. . . ." I have found absolutely no evidence of the existence of any "spy system" whatever.[16]

Dalby reports further that although the previous agent issued barbed wire to Indians with large numbers of cattle, the current agent issues barbed wire only for fencing cultivated land and not for pastures. Agent Reynolds, Dalby's report concludes, has achieved great results:

> When he arrived there were practically no fences, but tangled barbed wire was to be found everywhere, and there were practically no houses except for the whites. Now not an Indian receives a ration, they are all living on their allotments, and most of them are doing some kind of farm work for themselves, and are for the most part self-supporting, and probably a majority of them are housed, though it must be confessed that many of them still prefer to live in the tepee beside the house and use the house for storage purposes merely. Practically all the agricultural lands are fenced; the tangled barbed wire has been cleaned up and that which could be straightened out has been used, while the balance has been disposed of. . . . Considering the natural disposition of the Indian, and the fact that all this has been done in five years, his achievement is certainly creditable.[17]

What do we learn from the hearing? Barbed wire is in demand on the reservation and thus can be used for bribes. The Crows, despite their "natural disposition," now have houses, although they may still prefer their tepees. Rather than blankets, some of them wear citizens' clothing. They employ barbed wire fences on the cultivated portions of their allotments, at least the allotments not lost to white settlers. And finally, the tangled barbed wire the agent found when he arrived has been either straightened or disposed of. While this last achievement is meant literally, it is a good metaphor for the ongoing conquest of Native Americans. Dangerous and culturally tangled, they have been removed to reservations, untangled in boarding schools, or eliminated.

Enclosing and Enclosed

Literature by Native Americans is rich with the straight lines and tangles of barbed wire that delineate western landscapes and that feature ominously in their histories. James Welch's novel *Winter in the Blood*, for instance, is the story of a troubled Blackfeet/Gros Ventre man who, as the novel begins, returns to his home on the Fort Belknap Reservation in Montana:

> On the hill behind the cabin, a rectangle of barbed wire held the graves of all the Earthboys. . . . The fence hummed in the sun behind my back. . . . Coming home was not easy anymore. . . . I dropped down on the other side of the highway, slid through the barbed-wire fence and began the last two miles home.[18]

The barbed wire is a necessary ranching tool, and this is a family of ranchers. It is also a family of self-destructive drinkers, and the narrator has a hangover and a swollen eye from an encounter with a white man in town. The fence he slides through on his way home is an ambiguous border. The barbed wire surrounding the graves of the Earthboys is ambiguous as well, facing both inward and outward. This is an enclosed plot to which the dead have been banished (much like the family plot surrounded by sheep fence in "Brokeback Mountain," a final prison for the ashes of a man who cut the normative fences of rural behavior). The fenced graveyard in Welch's novel is also reminiscent of the reservation to which the living have been removed. Alternatively, the site protected by the fence is a private sanctuary for the Earthboys, although the "rectangle of barbed wire" is not a comforting image. The graveyard fence hums in the sun behind the narrator's back, an animate presence like so many of the fences we have discussed earlier. What is the message of the humming? Memento mori? Earthboy solidarity? A simple song about steel wire and the sun's heat, soothing, perhaps, after the violent confrontation between humans in the town? Fences reappear throughout the novel, as they must in a story set between a ranch and a town, but barbed wire is mostly absent after the opening scene. Perhaps it has done its work already, establishing the themes of borders, of border crossings, of enclosures, and of what Jane Bennett calls "vibrant matter," all critical for the story to come.[19]

Chris Eyre's film *Skins* (2002, based on the novel by Adrian C. Louis) offers another example of meaningful, if fleeting, images of barbed wire in a Native American work. The film opens with a shot of the South Dakota Badlands. An image of twisted barbed wire in the foreground dissolves as the camera focuses on the landscape of the Pine Ridge Reservation behind the wire. The story that follows the shot of the fence is about alcoholism, poverty, and despair, about a tribal police officer frustrated with a legal and social system that has failed his brother who is dying of cirrhosis of the liver. The initial barbed wire, in our reading, is a metaphor for conquest, for the enclosure of Native Americans on reservations, for the oppressive economic conditions and alcoholism that control the Oglala Lakota characters in the film. The film ends with a gesture of revolt when the police officer protests the death of his brother by pouring red paint down George Washington's face on Mount Rushmore. The gesture is full of bravado and expressive of frustration at centuries of Native oppression at the hands of a European American government. Cutting the barbed wire fences that enclose the reservation would have been symbolically powerful as well.[20]

Sterlin Harjo's film *Barking Water* (2009, written and directed by Harjo) features an elderly Indian couple on a final road trip along the back roads of Oklahoma, roads lined, invariably, by barbed wire fences. They have had difficult lives, in part of their own making, and now Frankie, the husband, is dying. As they drive and reminisce, the barbed wire is everywhere, much of it littered with windblown trash. Again the barbed wire serves as a metaphor of enclosure, of economic and social oppression, and of the conditions of a difficult life. Native residents live in a landscape ordered and reordered by barbed wire. For the most part, the fences signify property that has largely failed them, property to which they have been consigned, or property from which they are excluded.[21]

Wind from an Enemy Sky

A set of encounters between the US government and Native Americans serves as the basis of D'Arcy McNickle's *Wind from an Enemy Sky*. A reservation was created for the Flathead Indians in Montana under the 1855 Treaty of Hellgate. Eighteen tribal leaders who thought the meeting was about establishing peace between them and the Blackfeet reluctantly signed the treaty. They had no sense they were surrendering their lands to the federal government.[22] The next step of a process meant to erode claims by Native Americans to the land was the 1872 Allotment Act: "An act to provide for the removal of the Flathead and other Indians from the Bitter-root Valley, in the territory of Montana."[23] Subsequent allotment acts resulted in more than half of reservation lands being sold to white people. In 1930, Rocky Mountain Power began to build the Salish-Kootenai Dam, a concrete gravity-arch dam built on river mile 72 of the Flathead River. The dam was completed in 1938, destroying a productive fishery and drowning the sacred falls of the river.[24] An enrolled member of the Flathead tribe and eventually an official with the Bureau of Indian Affairs, D'Arcy McNickle began writing *Wind from an Enemy Sky* in the 1930s with the new dam in mind. Additionally, he drew on the story of a medicine bundle given to a museum by the Hidatsa tribe in North Dakota in response to the wishes of a Christian cleric. McNickle himself succeeded in having the bundle returned from the museum, but after a year of destructive flooding the bundle was sent back the museum.[25]

Native American identity, one component of McNickle's novel asserts, is tied to a traditional way of living with the land. The novel employs barbed wire fences to differentiate between the wild country still inhabited by Native American traditionalists led by a man named Bull and the supposedly civilized and subsidized existence chosen by Bull's brother Henry Jim. The brothers stand on opposite sides of the fence, one on the outside and the other on the inside. In concert with the Allotment Acts that broke up tribal lands, the barbed wire fences enforce individual property rights that conflict with the migratory rights of the past:

> The old people could remember when the Little Elk valley, like the sky above it, had been free and open. . . . Then surveying parties came and drove wooden stakes in the ground. After that, strangers from across the mountains came. And after that, wire was stretched between posts and the open country was gone. Now the soft rolling hills and gentle swales were cut into squares, and the traveler could only go where roads or lanes allowed passage between fence lines.[26]

Adam Pell, a wealthy white man with what he sees as benevolent intentions, looks down at the fenced plain and realizes that the conditions he and his government have imposed on the Indians are not what he, at least, had hoped for:

> From the vantage point of the hilltop it was possible to see a large expanse of valley, all of it cut into squares and rectangles by fence lines. . . . Adam gestured at the landscape. . . . "Look at those fences! More than all the documents I read last winter about dividing up the

> common tribal land and giving each man, woman, and child a limited piece of his inheritance, this view is the most shattering."[27]

Pell's assessment is echoed by the agency doctor: "If it hadn't been for [Bull], more of this Indian land would be in white ownership. . . . If the others had followed [Henry Jim], . . . they would have been suckered out of everything."[28] Communal use of the land was the traditional way of life for the Little Elks; private ownership destroys that tradition. As noted previously, the barbed wire fences of the American West are both the justification and the instrument for conquest.

Pell's museum failed adequately to protect the medicine bundle dear to the Little Elks, and he has come to give them a valuable pre-Inca Peruvian statuette "as a way of restoring something of what they lost through my carelessness—but also for my part in transforming their world into an alien land of fences and diverted water courses."[29] There is, of course, no way to compensate for the spiritual and physical losses of the medicine bundle and of the land. Although the statuette is extremely valuable in the economy Pell inhabits, what would the Little Elks do with it? When they realize Pell no longer has the medicine bundle, the tragedy that has threatened throughout the novel ensues.

Identity is intimately related to language, and while barbed wire restructures the landscape and Native life, the English language acts as another tool of conquest. Bull, for instance, is not who his name suggests he is: "He was named Bull—that was the English form of it. But the words men speak never pass from one language to another without some loss of flavor and ultimate meaning." When the Indian agent Rafferty visits Henry Jim, who has chosen to die in his tepee rather than in his house, Henry Jim tells the story of his assimilation, and in the telling switches back and forth between his native language and English. Rafferty tries to assess the situation: "The problem is communication. . . . We do not speak to each other—and language is only a part of it. Perhaps it is intention, or purpose, the map of the mind we follow."[30] Boarding schools for the children, as with the Quillayute children kept from going home on weekends by Agent McGlinn, impose European American language and culture while strictly forbidding Native language, dress, and custom. The competing worldviews described as incompatible mind maps that impede communication are manifest by the fences whose presence dominates the novel. Barbed wire is a visible, physical tool to impose a strictly rectangular European American map of the mind and to destroy Native identity.

The barbed wire fences are boundaries that subsume "all diversity into a single epistemological [and thus colonial] framework," according to Shari Huhndorf and Scott Pratt, who, interestingly, read Bull's insistence on a traditional mind map as another inflexible framework.[31] The conflict between colonial and traditional maps, in this reading, is a straightforward power struggle that only one party can win. The novel ends with Bull and others at the agency expecting the return of a medicine bundle that no longer exists. The discussion must be translated back and forth—a communication problem—and the message is life shattering. Bull responds by killing Rafferty and Pell and is shot himself. "*That day, the cry of the plover was heard everywhere . . . Ke-ree, ke-ree, ke-ree. No meadowlarks sang, and the world fell apart.*"[32] "The bird that spoke Lakota" falls silent with the death of Bull.[33]

Federal law, boarding schools, the English language, and a history of conquest impose a colonial "map of the mind" on the Native Americans. That mind map translates physically into allotments controlled by barbed wire fences and owned, eventually, by white settlers. Assimilation of some sort may be the only option over the long run, the novel suggests, but in the short run Bull's defiance sustains his version of Native identity. Searching for less destructive solutions in his book *Native American Tribalism*, McNickle cites a study indicating that "Indians remain Indians not by refusing to accept change or to adapt to a changing environment, but by selecting out of available choices those alternatives that do not impose a substitute identity."[34] What kind of identity would this be? Clearly not Bull's uncompromising Native identity. Nor that of assimilated Henry Jim. Two of the novel's characters, Antoine and Son Child or The Boy, inhabit both worlds and speak both languages. In the end, after the tragedy, they remain a hybrid hope for a future. "As border crossers," Huhndorf and Pratt conclude, "they continually challenge the fixed nature of and relations between cultures pictured by . . . colonial maps." Their new maps are "aids rather than ends," dynamic rather than static.[35] Dynamic alternatives to fixed identities are interesting to other Native Americans as well.

Off the Reservation

In her book *Off the Reservation*, Paula Gunn Allen celebrates her own hybridity: Laguna Pueblo, Maronite Lebanese, and Celtic Scot. Her pueblo is characterized by hybridity as well: "Laguna has from its inception been a mixed blood place, and to that distinguishing characteristic it owes not only its prospering survival, but its active and vital participation in many phases of American Life." Like McNickle, Allen asks how Native Americans can retain their "'indianness' while participating in global society." Native thought, she writes, does not fit "the categories that have been devised to organize Western intellectual enterprise," yet it must employ Western conventions if it is to engage Western thought. So, "like the half-breed, hybrid, mixed-blood woman who has composed them, these essays resemble the oral tradition of the Laguna world and the essayist tradition of the orthographic academy by turns."[36] Allen celebrates Old Man Coyote and the grand seducer Kokopelli and mocks Old Man Missionary, who finds

> himself fenced in by his own barbed-wire conventions, boundaries have grown permeable, wide open spaces abound. Gee, it's almost like old times!
>
> Native civilizations have their boundaries drawn along lines that differ markedly from those that characterize other civilizations. Probably the tendency of my migration-conscious Pueblo mind to see borders as both liminal and transformational led me early on to choose "Don't Fence Me In" as one of my favorite songs.[37]

But how does one retain one's cultural identity while interacting with other cultures, especially with majority cultures? Homi Bhabha raises this question in *The Location of Culture*. Like McNickle and Allen, Bhabha argues that it can be productive for fixed tradition to give way to hybridity:

> The representation of difference must not be hastily read as the reflection of pre-given ethnic or cultural traits set in the fixed tablet of tradition. The social articulation of difference, from the minority perspective, is a complex, on-going negotiation that seeks to authorize cultural hybridities that emerge in moments of historical trans-formation.[38]

What sorts of historical transformation have required this kind of negotiation for Allen's Laguna pueblo (a pueblo also at the heart of Leslie Marmon Silko's novel *Ceremony*)? The introduction of the railroad and of barbed wire were identity-challenging transformations. Reinforcing Allen's characterization of the Lagunas, Richard Frost contrasts the Santo Domingo and Laguna pueblos that, he writes, represented

> the pueblo extremes of reaction to the invasion of the Atchison, Topeka and Santa Fe. Santo Domingo sustained and intensified its conservatism, integrity, enclavement, cultural isolation, and resentment of the dominant society and its rising imperialism; Laguna strengthened its progressive response in education, jobs, and government, adjusting to modern conditions by compromising a preconceived pueblo identity and finding new opportunities in alien pressures.[39]

The Laguna response Frost describes here is a movement toward a cultural hybridity, as opposed to the Santo Domingo insistence on "the fixed tablet of tradition."

When, as New Mexico law required, the railroad fenced its right-of-way through Santo Domingo, the Natives protested. The fences split their land, they argued, prevented crossing except at given points, and when livestock strayed into the enclosed area, they could not find their way out and faced slaughter. "Underneath this was a profound belief of the Keresan-speaking Pueblos that fencing was a violation of mother earth and the common dependence of life on its surface." Santa Domingo did not even fence its own boundaries and thus suffered incursions from neighboring livestock. "The Assistant Commissioner remarked years later that if their land had belonged to the Laguna Indians 'they would have had it fenced long ago.'" In contrast to the Santa Domingans, the Lagunas prospered, Frost writes, by consolidating their allotments and reclaiming extensive tribal landholdings. Unfortunately, their growing herds contributed to overgrazing. The commissioner tried to draw on the War Department's surplus barbed wire in 1922 but was rebuffed. By 1931, when the Laguna finally fenced their land, the range was seriously overgrazed.[40]

Ceremony

Like Allen, and like Tayo, the protagonist of her novel *Ceremony*, Leslie Marmon Silko is of mixed parentage. She grew up in the Laguna pueblo next door to her Laguna great-grandmother. Silko began writing a short story that became the novel *Ceremony* while living on an Alaskan island that felt claustrophobic to a desert dweller. "I had a terrible kind of depression," she said, "and that was before I really got started writing. I finally wrote this short story. . . . It got me out of this terrible kind of depression. So I literally wrote *Ceremony* to save my life."[41]

Tayo, to save a life psychologically shattered by his experiences in World War II, some of which he spent behind barbed wire in a Japanese prisoner-of-war camp, makes his way through the events and landscapes of the novel while engaged in an improvised healing ceremony that involves a rich web of stories. *Ceremony* begins with several poems, the second of which is about the stories that are themselves ceremonies:

I will tell you something about stories,
[he said]
They aren't just entertainment.
Don't be fooled.
They are all we have, you see,
All we have to fight off
Illness and death.
. . . .
Their evil is mighty
But it can't stand up to our stories.
So they try to destroy the stories.[42]

These stories are both traditional and contemporary, and some of the stories that defend or defile these Laguna Indians have to do with barbed wire. Like the hybrid characters of the novel, barbed wire is a tool with sometimes contradictory capabilities. Barbed wire is a commonly used fence for cattle; it denies Indians and Mexicans access to land that was once theirs; and it works as a metaphor for the stories that demean and control Native Americans.

Well schooled by his uncle Josiah, Tayo always carries a pair of fencing pliers. According to Josiah the pliers are the second most important tool for a rancher (after a rope): "Hundreds of miles of barbed-wire fence marked boundaries and kept the cattle and horses from wandering. Josiah taught him to watch for lose strands of wire and breaks in the fence; he taught Tayo how to mend them before any livestock strayed off reservation land." This is Laguna pragmatism at work. Ranching requires fences, and the fences must be tended. This strand of the barbed wire story, as we have seen elsewhere, is also a vehicle for a broader discussion about human identity. Josiah tells Tayo that humans, as opposed to many animals, resist what they see outside themselves, that they refuse to accept a fence as a given. Animals, in contrast, become "part of the wind. . . . So they moved with the snow, became part of the snowstorm which drifted up against the trees and fences." Tayo responds with a skeptical question: "[and] when they died, frozen solid against a fence?" While discussing human and animal nature, Josiah and Tayo are following Josiah's spotted Mexican cattle, a hybrid breed less passive than the one that accepts death against a fence. The strong-willed hybrid cattle have broken through the barbed wire fence, and Tayo asks, "What if they just keep going, you know, crossing fences all the way back to Mexico?" Josiah is not worried, and in time "the Mexican cattle settled down and moved more slowly, but they still had little regard for fences."[43] These cattle, like Tayo, are a mixed breed adapted to the difficult conditions of the region; they have little respect for limiting, defining boundaries.

Despite Josiah's optimism about human ingenuity, Tayo has seen fences that cannot be breached. As a prisoner of the Japanese, Tayo watched a soldier kill his cousin Rocky, and then, when they reached the prison camp, "behind the barbed wire enclosed in many more layers of barbed wire, Tayo thought he saw the tall soldier come each day to stand beside the guard at the south fence and stare for a long time in his direction."[44] The gaze of the tall soldier and the many-layered barbed wire enclosure are reminiscent of the panoptical "argus eyes" of the Glidden advertisement discussed previously.[45] Memories of the vigilant guard and the barbed wire enclosure, of the death of his cousin, of the Japanese soldiers he was supposed to kill but who reminded him of his uncle Josiah, bedevil Tayo throughout the novel. To escape the debilitating "witchery" of these memories, Tayo seeks help from Ku'oosh, a medicine man who then sends him to Betonie, an unorthodox Navajo medicine man who has developed hybrid ceremonies for a new world: "After the white people came, elements in this world began to shift; and it became necessary to create new ceremonies."[46] This is a prime example of Bhabha's cultural hybridities emerging in "moments of historical trans-formation."

Like the prisoners in the Japanese camp, Mount Taylor, a sacred mountain taken from the Lagunas and relentlessly exploited by white ranchers and loggers, is enclosed by barbed wire. In the mountain's case, the wire faces outward to keep the Laguna from traditional hunting grounds, disturbing "the balance of the world."[47] Tayo follows old Betonie's vision up onto the mountain in search of the Mexican cattle. The first barbed wire fence is easily breached, and Tayo keeps searching:

> [Tayo] rode miles across dry lake flats and over rocky *cerros* until he came to a high fence of heavy-gauge steel mesh with three strands of barbed wire across the top. It was a fence that could hold the spotted

> cattle. The white man, Floyd Lee, called it a wolf-proof fence; but he had poisoned and shot all the wolves in the hills, and the people knew what the fence was for: a thousand dollars a mile to keep Indians and Mexicans out; a thousand dollars a mile to lock the mountain in steel wire, to make the land his.[48]

Tayo remembers that fencing pliers can also be useful, according to Josiah, when "a fence might get in your way. Josiah had nodded toward Mount Taylor when he said it."[49] And here the stories of the fence as a rancher's tool and of Mexican cattle and of humans who resist enclosure and of the prison camp and of Native American identity in the context of those stories are augmented by a single story told by the hegemonic white men:

> He was thinking about the cattle and how they had ended up on Floyd Lee's land. If he had seen the cattle on land-grant land or in some Acoma's corral, he wouldn't have hesitated to say "stolen." But something inside him made him hesitate to say it now that the cattle were on a white man's ranch. . . . Why did he hesitate to accuse a white man of stealing but not a Mexican or an Indian? . . . The Lie. He cut into the wire as if cutting away at the lie inside himself. The liars had fooled everyone, white people and Indians alike; as long as people believed the lies, they would never be able to see what had been done to them or what they were doing to each other.[50]

Tayo's fence cutting eliminates these kinds of borders. It enables the kind of crossing or mixing or transcending or overcoming that the structures of the ceremony encourage. In a sentence that perfectly describes the experience Tayo has while cutting the fence, Gerald Vizenor calls this kind of narrative work "survivance": "The simulations of survivance are heard and read stories that mediate and undermine the literature of dominance."[51]

Other kinds of borders, however, can be life saving. Tayo's growing sense for the pattern or constellation Betonie laid out for him is almost lost when he is thrown from his horse and knocked out. When he wakes, he feels the pull of death and explores the "thin walls, weak sutures of spindle bones above the ear for thresholds. He knew if he left his skull unguarded, if he let himself sleep, it would happen: the resistance would leak out and take with it all barriers, all boundaries; he would seep into the earth."[52] Merging with the earth would not be a tragedy. As Paula Gunn Allen has eloquently noted, "We are the land. . . . The land is not an image in our eyes but rather it is as truly an integral aspect of our being as we are of its being. And the integral nature of this fact continues beyond mortal dissolution of bodies—human, beast or plant."[53] It would, however, mark the end of Tayo's attempt to recover through the ceremony. The thresholds of Tayo's skull are simply boundary markers. Boundaries like the fence on the mountain, however, are enforcers of colonial power. The fences that control movement and the lies that control identity must be cut.

The final strand of the barbed wire stories occurs at night near a uranium mine. Until August 1945, guards, high barbed wire fences, and signs in Spanish and English had warned people to keep away from the site. Only remnants of those restrictions remain. Tayo finds his way to the mine while fleeing "friends" who intend to hurt him. He crawls through the fence, and there he finally sees "the pattern, the way all the stories fit together. . . . He was not crazy; he had never been crazy. He had only seen and heard the world as it always was: no boundaries, only transitions through all distances." The uranium mine, the laboratories of "Los Alamos, only a hundred miles northeast of him now, still surrounded by high electric fences," the bright light of the test blast his Old Grandma woke up to in the night, all intersect and he knows "there was no end to it; it knew no boundaries; and he had arrived at the point of convergence where the fate of all living things, and even the earth, had been laid." On the mountain, Tayo had grasped at thresholds to counteract the seduction of this unbounded convergence. Technological destruction of the land is witchery: "For more than two hundred years white people had worked to fill their emptiness; they tried to glut the hollowness with patriotic wars and with great technology and the wealth it brought." The wealth bought the mountain, which is fenced with barbed wire and lies. With their great technology, the "destroyers" have mined uranium and created bombs for patriotic wars and protected their secrets with barbed wire fences. The efficient fence thus becomes the servant of powers that aim to destroy all fences, all borders, all thresholds. Tayo has had this vast emptiness within himself until the ceremony began to provide constellations, patterns, and maps. Tayo's task this last night is to keep the story that is the ceremony "out of the reach of the destroyers for a few more hours."[54]

While Tayo watches them from a hiding place, his supposed friends Emo, Pinkie, and Leroy enact a vicious ritual. They haul Harley out of the car's trunk, disrobe him, burn his clothes, and torture him: "In the moonlight he could see Harley's body hanging from the fence, where they had tangled it upright between strands of barbed wire. . . . The screwdriver was slippery in his hands. It nauseated him to see Harley's body jerking and twitching in the sagging barbed wire, with hands and knives so greedy for human flesh." Tayo is sorely tempted to use the screwdriver "the way the witchery had wanted, savoring the yielding bone and remembrance as the steel ruptured the brain."[55] As was the case on the mountain when Tayo fought off the borderless lure of death by searching for the thresholds of his skull, here too the contours of the skull are the boundaries of reason. Tayo resists the temptation to violence in response to the witchery. Harley dies on the barbed wire.

Tayo's friends end up in early graves (or must leave the reservation in the case of Emo). The barbed wire used to torture Harley, the lies about white and Indian identity, and the technological control of nature so thoroughly enmesh these and other Indians that many of them end up violent and drunk and dead. Thanks to the ceremony, Tayo finds his way back to the kiva, where he tells Ku'oosh and the other old men what he has experienced. His new or renewed story is a narrative of ritual healing: "Every evil / which entangled him / was cut / to pieces."[56]

The connection made in the novel between barbed wire and nuclear weapons is reminiscent of the patterns we saw earlier in Abbey's and Meloy's works. Like them, *Ceremony* links the barbed fence to other technologies that destroy nature and annihilate cultures. Yet barbed wire persists as the fence, especially, of the American West.

Borderlands/La Frontera

The epigraph for Homi Bhabha's introduction to *The Location of Culture* comes from Martin Heidegger's "Building Dwelling Thinking": "A boundary is not that at which something stops but, as the Greeks recognized, the boundary is that from which something begins its presenting."[57] Bhabha argues,

> What is theoretically innovative, and politically crucial, is the need to think beyond narratives of originary and initial subjectivities and to focus on those moments or processes that are produced in the articulation of cultural differences. These "in-between" spaces provide the terrain for elaborating strategies of selfhood—singular or

communal—that initiate new signs of identity, and innovative sites of collaboration, and contestation, in the act of defining the idea of society itself.[58]

In her influential 1987 meditation on *Borderlands/La Frontera*, Gloria Anzaldúa explores ideas of identity in an extraordinary "in-between space" or interstice. She defines herself as the new mestiza, a creature of the dangerous borderlands between the United States and Mexico. Between genres (poetry, essay, history, memoir), between languages (English and various forms of Spanish), between cultures (Indian, Mexican, white), Anzaldúa chooses barbed wire as a metaphor for where and how she lives in the borderlands.

> I press my hand to the steel curtain—
> chainlink fence crowned with rolled barbed wire—
> rippling from the sea where Tijuana touches San Diego
> unrolling over mountains
> and plains
> and deserts,
> this "Tortilla Curtain" turning into *el rio Grande*
> flowing down to the flatlands
> of the Magic Valley of South Texas
> its mouth emptying into the Gulf.
>
> 1,950 mile-long open wound
> dividing a *pueblo*, a culture,
> running down the length of my body,
> staking fence rods in my flesh,
> splits me splits me
> me raja me raja
>
> This is my home
> this thin edge of
> barbwire.[59]

The barbed wire that stretches from the Pacific Ocean to the Gulf of Mexico is a tool of oppression, rusted witness to the years since the United States took over vast stretches of Mexico at the end of the Mexican-American War.[60] The idea of the thin edge of barbed wire that stakes and splits flesh as home is unthinkable. Anzaldúa thinks it, teases out the implications, embraces the fence that splits her. She moves from poetry to prose to think about border identities:

> The U.S.-Mexican border *es una herida abierta* where the Third World grates against the first and bleeds. And before a scab forms it hemorrhages again, the lifeblood of two worlds merging to form a third country—a border culture. Borders are set up to define the places that are safe and unsafe, to distinguish *us* from *them*. A border is a dividing line, a narrow strip along a steep edge. A borderland is a vague and undetermined place created by the emotional residue of an unnatural boundary. It is in a constant state of transition. The prohibited and forbidden are its inhabitants. *Los atravesados* live here: the squint-eyed, the perverse, the queer, the troublesome, the mongrel, the mulato, the half-breed, the half dead; in short, those who cross over, pass over, or go through the confines of the "normal."[61]

If borders are set up to divide us from them, and if she refuses that separation, then Anzaldúa must inhabit the border. She must claim the very machinery of violence that establishes the border. If the barbed wire is her home, she becomes dangerous rather than endangered. This new myth is hybrid, powerful, and affirming.[62] The barbed wire has split me? Then I am the barbed wire fence.

But (and the poem approaches its end with a "but") fences that rip a person and a pueblo apart ought themselves to be blown down; and for this the old myth is ready with the goddess of the ocean, who punishes the arrogance of the white man who has strung the barbed wire.

> But the skin of the earth is seamless.
> The sea cannot be fenced,
> *El mar* does not stop at borders.
> To show the white man what she thought of his
> Arrogance,
> *Yemayá* blew that wire fence down.[63]

Barbed wire enclosed the territories of Native migration and shared resources just as surely as it enclosed the free range. Barbed wire takes its place in Native American and mestiza texts as an actual and metaphorical definer and destroyer of cultures, separating us and them, enforcing boundaries through hegemonic stories, and, positively, establishing boundaries that invite courageous and creative hybrid responses.[64]

Coda

In his essay "Hall of Fame," John D'Agata describes more than two dozen Halls of Fame as he thinks about "the ways in which we matter." Among them is the "Magic Hall of Fame" in St. Louis, the "Billiards Hall of Fame" in Brooklyn, and the "National Cowboy Hall of Fame & Western Heritage Museum of Freedom" in Oklahoma City. After the long, proud title of the Cowboy Hall of Fame, D'Agata offers three epigraphs. The first is from the Mayflower Compact, 1620; the second, from an 1845 magazine ("the fulfilment of our manifest destiny is to overspread the continent allotted by Providence for the free development of our yearly multiplying millions"); and the third, Will Rogers's 1924 claim that his "forefathers didn't come over on the Mayflower, but they did meet the boat." The rest of the essay consists of a simple list of barbed wire patents:

> Stover's Star
> Crandal's Zigzag
> Crandal's Alternating Barb Strip
> Schuman's Double Edged Saw Tooth Slat
> . . . .
> Haish's Horn Barb
> Kelley's Grooved Pin
> Mary's Barbell
> Dodge's Knobby Barb
> Brink's Three Point Long Plate
> Brook's Variation

There are 157 names in all, 157 testimonies of the role of barbed wire in the conquest of the American West and its peoples.[65]

Conclusion

> Last night, in the rain, some of the men climbed over
> The barbed-wire fence of the detention center. . . .
>
> What does barbed wire feel like when you grip it, as though
> It were a loaf of bread, or a pair of shoes?
>
> — Mary Oliver, "Rain"

Mary Oliver, as poets often do, asks a tactile question: "What does barbed wire feel like when you grip it?" The only precise answer is that it feels like barbed wire, but Oliver is interested in something beyond tactile exactitude. Her possible answers—a loaf of bread, a pair of shoes, and other possibilities that follow—are all similes, as her question about "what it is *like*" demands. The shift from feeling barbed wire in your grip to exploring what it feels like is a construction of meaning. Constructions and connections of that sort have been the focus of our study.

What does barbed wire mean? How did it acquire those meanings? And what do those meanings mean for us? Answering these questions led us to present-day DeKalb, Illinois, to nineteenth-century newspapers from across a rapidly changing nation, to state legislatures, and to the US Patent Office. Creative and sometimes disturbing advertisements found in Colorado, Wyoming, Illinois, and Massachusetts archives gave us access to the minds of sellers and buyers of barbed wire. The question of meaning also led us to films and works of literature where landscapes and cowboys of the West and their counterparts in the New West have identities established and contested through barbed wire fences. That was no surprise. We were surprised, however, to find labor organizers and preachers of ascetic gospels responding to a barbed instrument for arousing abject ecstasy and to discover Native Americans, their lands stolen by invading fence builders, finding hybrid solutions to a barbarous fence that separates the Same from the Other. The transmutation of barbed wire into metaphor turned out to be a richer and more varied topic than we had anticipated.

Memories of human interactions with things that were once common litter our language. Generally, as we find ourselves increasingly removed from the real thing, our metaphors grow murky or even confused, as was the sports commentator who claimed that the new running back was champing at the bit like an unbridled horse. Unlike many such things that have faded into history, barbed wire is still a vivid presence in our physical and psychological landscapes. We know what it is, and we know what it can mean. Barbed wire, as Bill Brown claims about things in general, has helped us "make or re-make ourselves, to organize our anxieties and affections, to sublimate our fears and shape our fantasies."[1] The fact that barbed wire is efficient because it is dangerous, or to put it another way, that it controls by causing pain, makes the fence an especially good source of metaphor. The meaning of barbed wire depends on context, and as we have seen, there are many possible contexts. We end our study with several images and accounts that reemphasize the depth and breadth of the idea of barbed wire.

A Glidden House display of hundreds of varieties of barbed wire claims that barbed wire is "The Invention That Shaped the Nation" (fig. 37). The modest farmhouse Joseph Glidden was living in when he invented and produced the prototype of his Glidden Coil borders on land Glidden donated to what is now Northern Illinois University. Barbed wire surely shaped the lives of Glidden and his neighbors and fellow inventors Isaac Ellwood and Jacob Haish, and it dominates the city of DeKalb to this day. But is the Glidden House's broad claim simply small-town hyperbole, or did it really shape the nation? We think it did and that it still does.

Barbed wire is an efficient, if also vexing ranching tool (fig. 38). David Lee ends his poem "Barbed Wire" with these colloquial lines:

> I've waste more damn time on w[i]re today
> than I have to lose
> bring them pliers here

Figure 37. "The Invention that Shaped the Nation." Glidden House display, DeKalb, Illinois. Photograph by the authors.

Figure 38. Ranch fence in Snow, Woodland Hills, Utah. Photograph by the authors.

let's cut this sonofabitch it don't matter where
we gone set here all day
won't never get this damn fence done[2]

In Lee's narrative poem two men working on a barbed wire fence are frustrated by a roll of wire that will not reveal its end piece. The speaker mentions that he knew a man back home who invented a thing that would solve that problem and sent it to Washington for a patent, but the government never sent back an agreement. The inventor had a brother, he remembers, who was out tormenting pigs with a pellet gun when a storm came up. He was running for home when he was caught between strands of a barbed wire fence and the gun discharged, blinding him, and then lightning struck the fence—"made him a eunuch." So, blind and impotent, he became a gospel preacher. (O'Connor's Hazel Motes, likewise, became a gospel preacher and only later blinded and tortured himself.) A poem about trying to find the end piece of a roll of barbed wire, even one that includes a story about a pig tormenter caught and punished by a barbed wire fence, remains in the realm of routine ranch work: barbed wire will "tear the hide right off / your hands you seen them gloves?"[3] The poem could have been about shoeing horses or cutting hay, and, in fact, Lee's book includes poems about loading a boar, salvaging grain, digging postholes, and so on. What does barbed wire mean? It means work and requires gloves.

Like Lee's poem in that it depicts ranch work, but also about recording the work with artistic intent, Bruce Nauman's film of himself digging postholes and setting the posts for a barbed wire fence—*Setting a Good Corner (Allegory & Metaphor)*—as noted previously, explicitly asks a viewer to consider aspects of Western identity.[4] The Fool Creek Flat sign discussed earlier is another example of barbed wire as a signifier of identity (see fig. 2). The ranchers who populate novels of the Old West and the New West are unimaginable without the barbed wire fences that take away their free range, claim ranchland for them, control their cattle and horses, and embody their hopes for taut and intimate fences to ward off whatever they most fear.

The purpose of barbed wire is control. Because it exerts that control through nearly invisible wires twisted around sharp barbs, it can prove fatal. Incidents such as the one in which a pronghorn antelope died after becoming entangled

Figure 39. Pronghorn antelope hanging on barbed wire fence. Photograph by Benjamin Abbott.

in barbed wire fueled the intense debate we have chronicled in newspapers and before legislatures (fig. 39).

In ranching country, environmentalists may find the crucifixion of coyotes a warning to themselves, as did Amy Irvine in *Trespass*. Protesters at nuclear sites in the West are sure to encounter barbed wire–enforced boundaries. Terry Tempest Williams, whose mother died of breast cancer with likely origins in the aboveground nuclear testing in Nevada, writes about protesting in the contaminated town of Mercury, Nevada, a town they must reach by crossing a barbed wire border. "Under the cover of darkness, ten women slipped under a barbed-wire fence and entered the contaminated country." One vicious technology protects another. "They were trespassing."[5] And as they had planned, they were arrested. There is much to be gained by slipping under a barbed wire fence; there is meaning in crossing forbidding borders.

Barbed wire figured large in the European American conquest of the American West. It was used to establish ownership as it was imposed on Native

Figure 40. Fence sign, Laguna Pueblo, New Mexico. Photograph by Jan Wellington.

landscapes and, metaphorically, on Native minds. Laguna Indians like Tayo in Silko's *Ceremony* both cut the wire and use the fence, establishing hybrid identities that allow them to flourish in a changing world. Figure 40, a 2016 photo from the Laguna Pueblo, is an ironic declaration that the fence of conquest can be inverted and used against the perpetrators of conquest.

Knowing we were interested in anything having to do with barbed wire, friends brought a barbed wire cross back from a trip through central Colorado (fig. 41). The strands of cruel wire stapled to the cruel cross lay meaning over meaning. The cross itself stands for the suffering endured by Jesus Christ, and the body of Christ has been variously depicted as it hangs on the cross, most gruesomely, perhaps, by Matthias Grünewald in his *Isenheim Altarpiece*. Here, cross lies on cross, barbed wire on wood. Barbed wire is a perfect scourge and several characters in the literature we have examined use barbed wire to reduplicate the suffering of Christ. In the American West where this cross was found, barbed wire lends a sharp, local flavor to a Christian story.

Figure 41. Barbed wire cross.
In authors' possession.

The Christian connection appears in William Faulkner's *A Fable* as well. Faulkner imagines an event during World War I when a French corporal, disgusted by the senseless slaughter in the trenches, convinces fellow soldiers to quit fighting. German soldiers lay down their arms as well. Peace is unacceptable for those whose careers and reputations and hierarchical positions depend on ongoing warfare, and they attempt to contain the mutiny, converting a prison compound "into a man-proof pen for the mutinied regiment by the addition of . . . barricades of electrified wire and searchlight towers and machine-gun platforms and pits and an elevated catwalk for guards; French sappers and service troops were still weaving more barricades and stringing more of the lethal wire to crown them." Signaled in part by the wire that "had an appearance gossamer and ephemeral like Christmas tinsel," the corporal himself is a Christ figure. There is no recourse, finally, other than to execute him for the pacifist mutiny, at which point barbed wire becomes his crown of thorns: "The plunge of the post had jammed it and its burden too into a tangled mass of old barbed wire, a strand of which had looped up and around the top of the post and the man's head."[6] The wire so abundant in the trenches throughout the novel is recast here as a symbol of violence transcended. Similarly, the logo for Amnesty International features a spiral of barbed wire around a lit candle, another symbol of superseded violence.

Figure 42. Abandoned fence, San Rafael Swell, Utah. Photograph by the authors.

Barbed wire fences set up systems of order. They exert control. The condition of a fence signals how strong the control is, how strict the order. An abandoned fence, stretched originally on a precarious cliff on the San Rafael Swell in central Utah, is a slack reminder of how tenuous control is, of the fate of all ordered systems (fig. 42). Mero, the octogenarian Exercycle rider in Annie Proulx's story "The Half-skinned Steer," can only remember the taut and intimate fences he built on the Wyoming ranch while his last systems of control collapse.

Less intimate are the fences erected for political reasons, but those too must be maintained. In his book *Imperium*, Ryszard Kapuscinski recounts his 1958 Trans-Siberian journey that begins with musings about borders: "How many victims, how much blood and suffering, are connected with this business of borders!" It is no wonder that major religions promise a paradise with no borders, he writes: "That which is most desired, awaited, and longed for by everyone is precisely this unconditional, total, absolute—boundlessness."[7] Boundlessness is exactly what the barbed wire–infested border between China and the Soviet Union negates:

> Barbed wire. Barbed-wire barriers are what you see first. They protrude out of the snow, over it—lines, trestles, fences of barbed wire. What extraordinary combinations, knots, billows, entire constructions of barbed wire clasping together the sky and the earth, clinging to every bit of frozen field, to the white landscape, to the icy horizon. On the face of it, this thorny, rapacious barrier stretching along the border seems like an absurd and surreal idea, for who would force his way through here. As far as the eye can see there is only a desert of snow, no roads, no people, and the snow is two meters high; taking a single step is impossible. And yet these walls of barbed wire have something to say to you, have something to communicate. They are saying: Be careful, you are crossing the border into a different world. You will not escape from here, you will not get away. It is a world of deadly seriousness, orders, and obedience. Learn to listen, learn humility, learn to occupy the least amount of space possible. Best mind your own business. Best be silent. Best not ask questions.
>
> The barbed wire instructs you the whole time the train is rolling toward the station; it imprints upon your mind everything which you should from now on remember, and it does so relentlessly, but after all it is for your own good that it pounds into your head the long litany of limitations, prohibitions, and instructions.[8]

Border control, Kapuscinski finds, is both physical and mental. And barbed wire is a master teacher.

Planning a return trip in 1989, Kapuscinski pictures the immense space of the Soviet Union and the forty-two-thousand-kilometer border that encompasses it. Where technically possible, the borders are defined by barbed wire, wire that rapidly deteriorates in the dreadful climate, and Kapuscinski's mind turns to the immense task of erecting and replacing the fences: "One can assume that a significant portion of the Soviet metallurgical industry is devoted to producing barbed wire." Besides the external borders, he thinks, there are the camps of the gulag archipelago, what he later calls "a barbed net of destruction from which a man, once having fallen into it, could not extricate himself." "If one were to multiply all this by the number of years the

Soviet government has been in existence," he concludes, "it would be easy to see why, in the shops of Smolensk or Omsk, one can buy neither a hoe nor a hammer, never mind a knife or a spoon: such things could simply not be produced, since the necessary raw materials were used up in the manufacture of barbed wire."[9]

The scale of usage Kapuscinski describes cannot possibly have crossed the mind of Jacob Haish, Isaac Ellwood, or Joseph Glidden, but the Washburn & Moen Company already had some notion of the potential when it depicted barbed wire armor controlling humans at a Boston baseball field (see fig. 14). Totalitarian states like the Soviet Union come readily to mind when one thinks about border fences. Peter Leibing's photo of the East German soldier jumping tangled barbed wire in 1961 during the construction of the Berlin Wall is one of the best-known images of the Cold War. Sent to guard the border on the third day of the construction, nineteen-year-old Hans Conrad Schuman leaped over a roll of barbed wire, dropped his submachine gun, and was driven away by West Berlin police.

Closer to home, a painting done by a Japanese American during World War II is a reminder that Americans are also intimately acquainted with this coercive aspect of the perfect fence. Chiura Obata's *Regulations* is a response to his forced internment in Utah's Topaz Mountain camp with other American citizens of Japanese descent during the Second World War (fig. 43). His painting asserts the relationship between actual physical restraint and restraint by regulation by contrasting vertical rows of Japanese-language regulations with horizontal lines of barbed wire. As noted previously, the leader of the sect of barbed wire worshippers in Brian Evenson's story "Contagion" insists that "words must serve in the stead of a fence." Obata depicts both fence and words, understanding that without the regulations prescribed by President Roosevelt's Executive Order 9066, there would have been no camps surrounded by barbed wire, and without the barbed wire fences, the order could not have been enforced.[10]

"Don't call me babe" is her signature line, and you ignore Barbara Kopetski (a.k.a. Barb Wire) at considerable risk.[11] The heroine of John Arcudi's

Figure 43. *Regulations*, painting by Chiura Obata. From Chiura Obata, *Topaz Moon: Chiura Obata's Art of the Internment*, edited by Kimi Kodani Hill (Berkeley, CA: Heyday Books, 2000). Courtesy Heyday Books.

comic-book series—*Barb Wire, Steel Harbor's Babe on Wheels*—is as tough as barbed wire. She owns a bar called Hammerhead and works as a bounty hunter and sometime peacekeeper in the bombed-out town of Steel Harbor. She wears leather chaps and sports a tattoo of barbed wire on her upper left arm. She rides an Indian 841 motorcycle rather than a horse, and her weapons would reduce the O. K. Corral to splinters. This incarnation of the mythic Western hero is a woman, which is a bit surprising, but the name "Barb Wire" anchors her firmly in the tradition. "When you call me that, smile," taunts the tigerlike Virginian. "Don't call me babe," challenges Barb Wire.[12]

Barbed wire has come to mean toughness. It means danger. It means death. And it means the worst trouble a songwriter can imagine:

I walked 47 miles of barbed wire,
Used a cobra snake for a neck tie.
Got a brand new house on the roadside,
Made out of rattlesnake hide.
I got a brand new chimney made on top,
Made out of human skulls.
Now come on darling let's take a little walk, tell me,
Who do you love,
Who do you love, Who do you love, Who do you love.[13]

Bo Diddley's gravel-throated "Who Do You Love?" would have made no sense had he walked "47 miles" of picket fence. Barbed wire, cobras, rattlesnakes, human skulls. That's love. And it is music as well.

Figure 44. Salvador Dalí sketches Harpo Marx playing the barbed wire harp Dalí had made for Marx. Courtesy Fundació Gala-Salvador Dalí.

In 1936 Salvador Dalí sent his friend Harpo Marx a harp with barbed wire strings and spoons for tuning knobs. Marx sent back a photo of himself playing the harp with bandaged fingers. Figure 44 is a photo from Dalí's 1937 visit to California where he sketched Marx playing the musical conundrum.[14] What does barbed wire mean? Pain is the short answer, and that makes barbed wire a perfect medium for Dalí, who stretches it where it least belongs, where the pain will be the ludicrous point.

If setting barbed wire in an incongruous context produces humor, installing it in a danger zone is no joke. In a long poem about travel in Israel, Israeli poet Yehuda Amichai finds ruined vineyards at the site where he first knew love, and at a second site he finds a dried up wellspring. He then travels at night through unexpected rain to Ein Yahav in the Aravah Desert:

> A night journey to Ein Yahav in the Aravah,
> a journey in rain. Yes, rain.
> There I met people growing dates.
> There I saw tamarisk trees and trees of illusion.
> There I saw a barbed hope, like barbed wire.
> And I said to myself, "It's true: Hope must be like
> barbed wire to protect us from despair.
> Hope must be like a minefield."[15]

Amichai finds settlers growing date palms and tamarisk trees, trees of illusion, he calls them—an illusion because of the desert, perhaps, or more likely because of the dangers associated with Jewish settlement in Palestine. Ein Yahav was established as an agricultural station in 1950 and then converted into a Nahal settlement to defend against potential Arab invasions.[16] Hope is necessary for inhabitants of a desert village, even more essential for dwellers at an outpost on a contested border. And thus the need for barbed hope, hope formidable enough to keep despair at bay. As the Glidden advertisement for barbed wire claimed: "It prevents the 'ins' from being 'outs'; and the 'outs' from being 'ins.'" Who should be in? Who should be out? These political questions, even today, are answered tenuously (and viciously) by walls and barbed wire fences until more reasonable answers are found.

Wislawa Szymborska's "Starvation Camp near Jaslo" reminds us that barbed wire is at the heart of every war. The poet imagines every tangible and visible thing in the Nazi prison camp as the thousands of starving prisoners might have—as food. Even the barbed wire that imprisons the inmates becomes an object related to food:

> A man swayed
> On a grill of barbed wire.
> Some sang, with dirt in their mouths. *That lovely song*
> *About war hitting you straight in the heart.*[17]

There are no lovely songs about barbed wire.

NOTES

Introduction

1. Jack London, *The Road* (New York: MacMillan, 1907), 187–188.

2. Google Ngram, "Barbed Wire," accessed March 12, 2017, https://books.google.com/ngrams/graph?content=barbed+wire&year_start=1870&year_end=2016&corpus=15&smoothing=3&share=&direct_url=t1%3B%2Cbarbed%20wire%3B%2Cc0.

3. Leo Marx, *The Machine in the Garden* (New York: Oxford University Press, 2000), 3.

4. Ibid., 181, 191, 191, 195, 206.

5. Ibid, 178.

6. Ibid., 164.

7. Jean-Jacques Rousseau, *The Major Political Writings of Jean-Jacques Rousseau: The Two "Discourses" and the "Social Contract,"* ed. and trans. John T. Scott (2012; repr., Chicago: University of Chicago Press, 2014), 91.

8. H. G. Wells, *The Wonderful Visit* (New York: E. P. Dutton, 1895), 185.

9. Ibid., 205–207.

10. Thomas Oles, *Walls: Enclosure and Ethics in the Modern Landscape* (Chicago: University of Chicago Press, 2015), 9, 70, 79, 82, 83, 94.

11. *Steel Barb Fencing, the World's Fence, Everywhere Adapted to All the Uses of Husbandry in All Climates and under All Exposures, "the Perfect Fence"* (Worcester, MA: Snow, Woodman, 1880), 8. Alan Krell also reads this and other statements in the manual as proof of colonizing intent. See *The Devil's Rope: A Cultural History of Barbed Wire* (London: Reaktion Books, 2002), 12.

12. Letter from the Secretary of the Interior, Transmitting, in Answer to Senate Resolution of February 14, Letter of the Commissioner of the General Land Office, on the Subject of Unauthorized Fencing of Public Lands, Exec. Doc. No. 127, 48th Cong., 1st sess., March 14, 1884, 2.

13. Ibid., 2–3.

14. Ibid., 22.

15. Ibid., 21.

16. Public Lands. Unlawful Inclosures or Occupancy; Obstructing Settlement or Transit, U.S.C. 1061, Title 43 (1885).

17. "Proclamation No. 271—Ordering the Immediate Removal of Any and Every Unlawful Enclosure of Public Land," August 7, 1885, The American Presidency Project, http://www.presidency.ucsb.edu/ws/?pid=71889.

Part 1

1. "DeKalb, Illinois," City of DeKalb website, accessed February 20, 2017, http://www.cityofdekalb.com/854/History-of-DeKalb.

2. Henry M. Rose, Improvement in Wire Fences, US Patent 138,763, filed December 31, 1872, issued May 13, 1873.

3. Henry D. McCallum and Frances T. McCallum, *The Wire That Fenced the West* (Norman: University of Oklahoma Press, 1965), 29–45.

4. *The DeKalb County Manufacturer, with Illustrations and Historical Sketches of DeKalb County Inventions* (DeKalb, IL: Chronicle Press and Bindery, 1882), Earl Hayter Collection, Box 2, Folder 24, American Heritage Center, University of Wyoming, Laramie (hereafter AHC).

5. Ibid., 16.

6. Ibid., 17.

7. Ibid.

8. Ibid., 17, 18. Contrary to Ellwood's statement, the 1880s are rife with public debate about the nature and meaning of barbed fencing. As discussed later, barbed wire undergoes design changes to address the public outcry.

9. Ibid., 21.

10. Ibid., 22.

11. Ibid., 24, 25.

12. Henry and Frances McCallum, for instance, note that Haish was called "the irascible German." See *The Wire That Fenced the West*, 88.

13. I. L. Ellwood, letter to H. B. Sanborn, November 6, 1876, American Steel & Wire Company Collection, Baker Business Historical Collection, Harvard University (hereafter BBHC). See A. K. Stiles, president, Thorn Wire Hedge Company, flier "The Eccentricities of Jacob Haish," which refers to Haish's literature as "funny," BBHC.

14. I. L. Ellwood, letter to Washburn & Moen Co., April 30, 1881, BBHC.

15. Ibid., June 8, 1881, BBHC.

16. Ibid., June?, 1881, BBHC.

17. Ibid., June 27, 1881, BBHC.

18. Ibid., June?, 1881, BBHC.

19. "Smoke," *Barb Fence Regulator. Senior*, vol. 6 (DeKalb, IL: Jacob Haish, n.d.).

20. The "trivial thing" was the listing of customers' names. See John Gedge and William Boley, "History of the Manufacture of Barbed Wire," comp. G. L. Meaker and L. C. Bailey, for the American Steel & Wire Company of New Jersey, 1913, cited in McCallum and McCallum, *The Wire That Fenced the West*, 94.

21. I. L. Ellwood, letter to Washburn & Moen Co., received by Washburn August 23, 1881, BBHC.

22. Jacob Haish, *A Reminiscent Chapter from the Unwritten History of Barb Wire prior to and Immediately following the Celebrated Decision of Judge Blodgett,* December 15, 1880 (N.p., 188?), 2.

23. Ibid., 18.

24. Justia US Supreme Court, "The Barbed Wire Patent," 143 US 275 (1892), 143 US 292, https://supreme.justia.com/cases/federal/us/143/275/case.html.

25. Arjun Appadurai, ed., *The Social Life of Things: Commodities in Cultural Perspective* (Cambridge: Cambridge University Press, 1986), 3, 6.

26. Jane Bennett, *Vibrant Matter: A Political Ecology of Things* (Durham, NC: Duke University Press, 2010), 122.

27. Thomas Schlereth, *Material Culture Studies in America: An Anthology Selected, Arranged, & Introduced by Thomas Schlereth* (New York: Altamira Press, 1991), 3.

28. James Deetz, *In Small Things Forgotten: The Archaeology of Early North America Life*. Exp. and rev. (New York: Anchor Books, 1996), 35.

29. Sterling Evans, *Bound in Twine: The History and Ecology of the Henequen-Wheat Complex for Mexico and the American and Canadian Plains, 1880–1950* (College Station: Texas A&M University Press, 2007), 240.

30. Gregory T. Cushman, *Guano and the Opening of the Pacific World: A Global Ecological History* (New York: Cambridge University Press, 2013), xiii.

31. Witold Rybczynski, *One Good Turn: A Natural History of the Screwdriver and the Screw* (New York: Scribner, 2000).

32. Jack Glover, *"Bobbed" Wire: An Illustrated Guide to the Identification and Classification of Barbed Wire* (Wichita Falls, TX: Terry Bros., 1966); R. J. Thurgood, *The Complete Encyclopedia of Barbed Wire* (Paducah, KY: Collector Books, 1972); and Harold Hagemeier, *The Barbed Wire Identification Encyclopedia*, 5th ed. (Kearney, NE: Morris Press, 2010)—this edition was followed by the *Supplement to Edition Five of Barbed Wire Identification Encyclopedia,* 5th ptg. (Kearney, NE: Morris Press, 2012).

33. See Arthur G. Warren, *Barbed-Wire—Who Invented It? History of Early Development and Litigation* (N.p., 1926); and Arthur G. Warren, *History of the Manufacture of Barbed Wire Fencing: Abstract from the Files of the Industrial Museum, American Steel & Wire Company* (Worcester, MA: American Steel & Wire, 1924).

34. Walter Prescott Webb, *The Great Plains* (Lincoln: University of Nebraska Press, 1931), 312, 271, 312, 317.

35. Earl W. Hayter, "Barbed Wire Fencing—a Prairie Invention: Its Rise and Influence in the Western States," *Agricultural History* 13, no. 4 (1939): 207, 205–206.

36. Earl W. Hayter, "The Fencing of Western Railways," *Agricultural History* 19, no. 3 (1945): 164–166.

37. Wayne Gard, "The Fence-Cutters," *Southwestern Historical Quarterly* 51, no. 1 (1947): 15.

38. C. Boone McClure, "History of the Manufacture of Barbed Wire," *Panhandle Plains Historical Review*, 23(1958):1–114.

39. Joseph M. McFadden, "Monopoly in Barbed Wire: The Formation of the American Steel and Wire Company," *Business History Review* 52, no. 4 (1978): 467–470.

40. Kathryn Byrd's master's thesis, "Barbed Wire: Its Origin and Effects" (Southwest Texas State Teachers College, 1944), is largely a secondary-source synthesis that demonstrates a growing amount of literature on the subject of barbed wire fencing in the West.

41. McCallum and McCallum, *The Wire That Fenced the West*, 68–72.

42. Reviel Netz, *Barbed Wire: An Ecology of Modernity* (Middletown, CT: Wesleyan University Press, 2004), xii.

43. Ibid., 233. One of the earliest military applications of barbed wire occurred during the Spanish-American War in 1898, when it was used for barricades. Teddy Roosevelt and his Rough Riders became some of the first wire cutters in wartime at the Battle of Kettle Hill. See Charles Herner, *The Arizona Rough Riders* (Tucson: University of Arizona Press, 1970), 140.

44. Olivier Razac, *Barbed Wire: A Political History* (New York: New Press, 2003), 21–22, 106.

45. Krell, *The Devil's Rope*, 147. More recently Joanne S. Liu published a popular history called *Barbed Wire: The Fence That Changed the West* (Missoula, MT: Mountain Press, 2009). It is a quick retelling of the story told more thoroughly and with greater insight by Henry and Frances McCallum.

Chapter 1

1. This account is included in McCallum and McCallum, *The Wire That Fenced the West*, 47. Grenniger filed no patent for his attachment, but the McCallums conclude it was one of the "earliest known examples of protective fencing armed with spikes"(48). It is not known if Grenniger's demise was directly related to his continued use of an armored fence.

2. Ibid., 50.

3. *Los Angeles Daily Herald* 4, no. 36 (May 7, 1875): 3.

4. *Wichita City Eagle* (Wichita, Sedgwick County, KS) 6, no. 40 (January 3, 1878): 4.

5. *Cultivator & Country Gentleman* 42, no. 1275 (July 5, 1877): 423.

6. *American Agriculturalist*, n.s. 394, 38, no. 11 (November 1879): 457.

7. See, for example, *Steel Barb Fencing*, front cover.

8. *Los Angeles Daily Herald* 4, no. 36 (May 7, 1875): 3.

9. Ibid., no. 43 (May 15, 1875): 4.

10. *Waco Daily Examiner* (Waco, TX) 4, no. 85 (December 19, 1875): 3.

11. *Emporia News* (Emporia, KS) 20, no. 51 (December 21, 1877): 2.

12. *Elk County Advocate* (Ridgway, Elk County, PA) 8, no. 36 (October 24, 1878): 4.

13. *Cultivator & Country Gentleman* 47, no. 1559 (December 14, 1882): 981–982.

14. *Butler Weekly Times* (Butler, MO) 5, no. 12 (February 21, 1883): 2.

15. *Phillipsburg Herald* (Phillipsburg, KS) 5, no. 32 (May 31, 1883): 6.

16. *Elk County Advocate* (Ridgway, Elk County, PA) 8, no. 36 (October 24, 1878): 4.

17. *Emporia News* (Emporia, KS) 20, no. 51 (December 21, 1877): 2.

18. *Arizona Sentinel* (Yuma, Arizona Territory) 6, no. 43 (February 2, 1878): 3.

19. *Cultivator & Country Gentleman* 42, no. 1287 (September 27, 1877): 616.

20. *Daily Globe* (St. Paul, MN) 2, no. 228 (August 30, 1879): 2.

21. *Iola Register* (Iola, Allen County, KS) 17, no. 19 (April 27, 1883): 5.

22. Ibid., no. 20 (May 4, 1883): 5.

23. *Century Magazine, Illustrated Monthly* 30, no. 1 (May 1885): 13.

24. *Rock Island Argus* (Rock Island, IL) 27 (April 19, 1878): 4.

25. Ibid., June 19, 1878, 4.

26. *Bismarck Tribune* (Bismarck, Dakota Territory) 7, no. 5 (June 28, 1879): 4.

27. *Weekly Arizona Miner* (Prescott, Arizona Territory) 15, no. 32 (July 26, 1878): 3; and *Middlebury Register* (Middlebury, VT) 44, no. 14 (July 4, 1879): 3.

28. *News and Citizen* (Morrisville and Hyde Park, VT) 10, no. 10 (May 25, 1882): 3.

29. *Iola Register* (Iola, Allen County, KS) 14, no. 41 (October 1, 1880): 1.

30. *Nebraska Advertiser* (Auburn, NE) 27, no. 1 (June 22, 1882): 4.

31. *Cultivator & Country Gentleman* 47, no. 1554 (November 16, 1882): 905.

32. *Dallas Daily Herald* 30, no. 75 (February 16, 1883): 2.

33. *Southern Cultivator and Dixie Farmer* 51, no. 6 (June 1893): 286.

34. "Outside Echoes: Beautifying a Prairie Home," *Press and Daily Dakotaian* (Yankton, Dakota Territory) 9, no. 65 (July 12, 1883): 1.

35. *Scribner's Magazine* 20, no. 3 (September 1896): 384.

36. *Iola Register* (Iola, Allen County, KS) 17, no. 21 (May 11, 1883): 8.

37. *Wheeling Daily Intelligencer* (Wheeling, WV) 3, no. 292 (August 11, 1882): 4.

38. *Daily Globe* (St. Paul, MN) 6, no. 150 (May 30, 1883): 5.

39. *Marshall County Republican* (Plymouth, IN) 22, no. 13 (March 14, 1878): 4.

40. *Northern Pacific Farmer* (Wadena, MN) 3, no. 25 (November 11, 1880): 1.

41. *St. Landry Democrat* (Opelousas, LA) 3, no. 47 (December 11, 1880): 2.

42. *Semi-Weekly Interior Journal* (Stanford, KY) 10, no. 510 (August 1, 1882): 2.

43. *Milan Exchange* (Milan, TN) 10, no. 8 (April 21, 1883): 5.

44. "Fort Collins," *Daily News* (Denver, CO), April 25, 1882, 3.

45. *Leavenworth Weekly Times* (Leavenworth, KS) 2, no. 19 (September 5, 1878): 1.

46. *Dodge City Times* (Dodge City, KS) 5, no. 2 (May 25, 1878): 1.

47. *Emporia News* (Emporia, KS) 21, no. 32 (August 9, 1878): 3.

48. *Weekly Kansas Chief* (Troy, KS), 22, no. 12 (September 5, 1878): 3. *Leavenworth Weekly Times* (Leavenworth, KS) 25, no. 19 (September 5, 1878): 1.

49. *Lexington Weekly Intelligencer* (Lexington, MO) 9, no. 5 (April 12, 1879): 2; no. 10 (May 17, 1879): 3.

50. *Lewiston Teller* (Lewiston, North Idaho) 3, no. 34 (June 6, 1879): 3.

51. *Daily Globe* (St. Paul, MN) 2, no. 224 (August 26, 1879): 2.

52. *New Orleans Daily Democrat* 4, no. 237 (August 12, 1879): 2.

53. *Emporia News* (Emporia, KS) 23, no. 39 (September 24, 1880): 1.

54. *San Saba News* (San Saba, TX) 6, no. 38 (April 21, 1880): 4. The screwworm plague remained a major concern to stock raisers well into the twentieth century.

55. *Mower County Transcript* (Austin, Mower County, MN) 9, no. 13 (June 29, 1876): 3.

56. *Burlington Weekly Free Press* (Burlington, VT), n.s. 25, 53, no. 11 (September 12, 1879): 1.

57. *Green-Mountain Freeman and Record and Farmer* (Montpelier, VT) 36, no. 42 (October 15, 1879): 4. This story, taken largely from the magazine *New England Farmer*, was reprinted two weeks later in the *Vermont Phoenix* (Brattleboro, VT) 46, no. 44 (October 31, 1879): 4.

58. *Lewiston Teller* (Lewiston, North Idaho) 3, no. 32 (May 23, 1879): 3.

59. *Emporia News* (Emporia, KS) 22, no. 20 (May 16, 1879): 3.

60. *Weekly Clarion* (Jackson, MS) 43, no. 38 (September 22, 1880): 4. Similar suggestions appeared in the *River Press* (Fort Benton, Montana Territory) 1, no. 2 (November 3, 1880): 2; *Red Cloud Chief* (Red Cloud, Webster County, NE) 8, no. 20 (December 23, 1880): 1; *Rocky Mountain Husbandman* (White Sulphur Springs, Montana Territory) 5, no. 14 (February 19, 1880): 1; and *Los Angeles Daily Herald* 15, no. 76 (April 29, 1881): 2.

61. *Jackson Standard* (Jackson, OH) 38, no. 32 (August 10, 1882): 2.

62. *Green-Mountain Freeman* (Montpelier, VT) 36, no. 18 (April 30, 1879): 2.

63. *New North-West* (Deer Lodge, Montana Territory) 13, no. 4 (July 22, 1881): 1.

64. *Emporia News* (Emporia, KS) 24, no. 32 (August 11, 1881): 4.

65. *New Ulm Weekly Review* (New Ulm, MN) 4, no. 28 (July 6, 1881): 2.

66. *Brenham Weekly Banner* (Brenham, Washington County, TX) 17, no. 23 (June 8, 1882): 1.

67. *The Times* (Owosso, MI), n.s., 2, no. 17, whole no. 217 (July 20, 1883): 8. The number of cattle deaths in this story ensured it was picked up almost verbatim in other outlets, including the *Prairie Farmer* 55, no. 30 (July 28, 1883): 468; *Sacramento Daily Record-Union* 50, no. 10 (September 26, 1883): 135, *Record* 28, no. 5140 (September 26, 1883): 6.

68. *National Tribune* (Washington, DC) 2, no. 14, whole no. 66 (November 16, 1882): 6.

69. *Democratic Northwest* (Napoleon, Henry County, OH) 31, no. 19 (July 5, 1883): 6.

70. *Prairie Farmer* 59, no. 7 (February 13, 1886): 110.

71. *Las Vegas Daily Gazette* (Las Vegas, New Mexico Territory) 5, no. 28 (May 13, 1883): 4.

72. *Evening Star* (Washington, DC) 62, no. 9549 (November 30, 1883): 1. Two years later, the National Tanners and Hide and Leather Dealers' Association met and adopted a resolution "condemning the use of barb-wire fences as cruel to animals and injurious to their hides." *New York Times*, 34, no. 10,621 (September 17, 1885): 2.

73. *Ottawa Free Trader* (Ottawa, IL) 6, no. 52 (July 29, 1876): 1.

74. *Weekly Kansas Chief* (Troy, KS) 69, no. 17 (October 10, 1878): 3.

75. *Weekly Democratic Statesman* (Austin, TX) 12, no. 18 (December 7, 1882): 1.

76. *New Ulm Weekly* (New Ulm, MN) 5, no. 11 (March 8, 1882): 3.

77. *St. Landry Democrat* (Opelousas, LA) 5, no. 13 (April 15, 1882): 3.

78. *Rock Island Argus* (Rock Island, IL) 30, no. 210 (June 24, 1882): 1.

79. *Barbour County Index* (Medicine Lodge, Barbour County, KS) 3, no. 16 (September 22, 1882): 6.

80. *Chariton Courier* (Keytesville, Chariton County, MO) 7, no. 46 (January 25, 1879):1. See also *Mower County Transcript* (Austin, Mower County, MN) 12, no. 20 (August 14, 1879): 3; *Leavenworth Weekly Times* (Leavenworth, KS) 26, no. 1284 (September 11, 1879): 6; *Iola Register* (Iola, Allen County, KS) 13, no. 46 (November 7, 1879): 1; *New Bloomfield, PA Times* (New Bloomfield, PA) 14, no. 16 (April 13, 1880): 4; *Weekly*

Democratic Statesman (Austin, TX) 10, no. 3 (August 19, 1880): 2; "A Victim of Barbed Wire," *Emporia News* (Emporia, KS) 24, no. 17 (April 29, 1881): 1; *Waco Daily Examiner* (Waco, TX) 15, no. 122 (May 9, 1882): 1; and *Lexington Weekly Intelligencer* (Lexington, MO) 12, no. 21 (July 29, 1882): 2.

81. "The Brutal Barb: A Woman Fearfully Cut Up by Being Thrown against a Barbed Fence," *Omaha Daily Bee* (Omaha, NE) 12, no. 108 (October 22, 1883): 6.

82. *Mower County Transcript* (Austin, Mower County, MN) 16, no. 14 (June 27, 1883): 1.

83. *Phillipsburg Herald* (Phillipsburg, KS) 6, no. 1 (November 15, 1883): 5.

84. *National Live-Stock Journal-Weekly Edition* (Chicago, IL) 4, no. 51 (December 18, 1888): 803.

85. *Chariton Courier* (Keytesville, Chariton County, MO) 8, no. 24 (August 23, 1879): 2.

86. *Belmont Chronicle* (St. Clairsville, OH) 22, no. 6 (February 16, 1882): 3. See also *Omaha Daily Bee* (Omaha, NE) 12, no. 144 (December 4, 1882): 6.

87. *Wichita City Eagle* (Wichita, Sedgwick County, KS) 11, no. 44 (January 25, 1883): 3. See also *Sacramento Daily Record-Union* (Sacramento, CA): *Record* 59, no. 10 (June 1, 1883): 3, *Union* 27, no. 5037 (June 1, 1883): 3; and *McCook Weekly Tribune* (McCook, Red Willow County, NE) 2, no. 29 (December 20, 1883): 2.

88. *Emporia News* (Emporia, KS) 24, no. 32 (August 11, 1881): 4.

89. Ibid. 23, no. 52 (December 24, 1880): 3.

90. "Climbed the Fence: Three Girls Escape from the Erring Woman's Home," *Chicago Daily Tribune* (Chicago, IL) 51, no. 207 (July 25, 1892): 3.

91. *Idaho Semi-Weekly World* (Idaho City, Idaho Territory) 8, no. 69 (November 6, 1883): 2.

92. *New Ulm Weekly Review* (New Ulm, Brown County, MN) 9, no. 27 (July 7, 1886): 7.

93. "Barbed Fencing," *Cultivator & Country Gentleman* 42, no. 1275 (July 5, 1877): 423.

94. *Cultivator & Country Gentleman* 43, no. 1308 (February 21, 1878):115.

95. Ibid., no. 1311 (March 14, 1878): 163–164, 164.

96. Ibid., no. 1317 (April 25, 1878): 261.

97. Ibid., no. 1324 (June 13, 1878): 376.

98. Ibid. 45, no. 1422 (April 29, 1880): 276.

99. Ibid., no. 1427 (June 3, 1880): 362.

100. Ibid., no. 1423 (May 6, 1880): 292.

101. Ibid., no. 1424 (May 13, 1880): 312.

102. Ibid., no. 1429 (June 17, 1880): 389.

103. Ibid., no. 1435 (July 29, 1880): 484.

104. Ibid., no. 1440 (September 2, 1880): 570.

105. *American Agriculturalist* 39, no. 1 (January 1880): 10.

106. Ibid., no. 4 (April 1880): 134.

107. Ibid., no. 3 (March 1880): 93. Despite its claims of numerous letters in opposition to barbed fencing, the *American Agriculturalist* published only one such letter,

which was lifted from the *Cultivator & Country Gentleman*. The letter in question was not from a fence user but from the formidable veterinary surgeon William Horne. Ibid., no. 6 (June 1880): 223. In the June 1880 issue, the editor noted, "Our first strong prejudice against Barbed-wire fences was gathered at Lincoln, Neb., and confirmed afterwards at Beloit, Wis., not far from Janesville. It was there that we first met with the Brinkerhoff steel strap, which, with the modified barbs we suggested several months ago, we still believe to be the most desirable fence." Ibid. 39, no. 6 (June 1880): 223.

108. Ibid., no. 3 (March 1880): 93.

109. Ibid., no. 5 (May 1880): 179, 180.

110. Ibid., no. 7 (July 1880): 262.

111. Ibid., no. 3 (March 1880): 93.

112. Ibid., no. 5 (May 1880): 179–180.

113. Ibid., no. 4 (April 1880): 134.

114. Ibid. 42, no. 2 (February 1883): 63.

115. *Lancaster Daily Intelligencer* (Lancaster, PA) 17, no. 210 (May 5, 1881): 4.

116. Ibid. 28, no. 166 (March 16, 1882): 4.

117. *The Meridional* (Abbeville, LA) 14, no. 3 (July 16, 1881): 2; *Weekly Register* (Point Pleasant, Mason County, WV) 20, no. 46 (December 28, 1881): 4; and *Rocky Mountain Husbandman* (White Sulphur Springs, Montana Territory) 8, no. 8 (January 4, 1883): 1.

118. *Daily Globe* (St. Paul, MN) 4, no. 168 (June 17, 1881): 1.

119. Ibid. 6, no. 218 (August 6, 1883): 4.

120. "Barberous Wire Fence," *News and Herald* (Winnsboro, SC) 4, no. 48 (April 20, 1880): 4.

121. *Weekly Arizona Miner* (Prescott, Arizona Territory) 18, no. 21 (May 20, 1881): 4.

122. *Worthington Advance* (Worthington, Nobles County, MN) 9, no. 42 (June 16, 1881): 3.

123. *Weekly Miner* (Butte, Montana Territory) 5, no. 268 (July 26, 1881): 8.

124. *Caledonian* (St. Johnsbury, VT) 47, no. 2404 (August 24, 1883): 2.

125. *Saline County Journal* (Salina, KS) 13, no. 36 (October 4, 1883): 3.

126. *Richmond Democrat* (Richmond, Ray County, MO) 9, no. 5 (January 27, 1881): 2.

127. *Mexico Weekly Ledger* (Mexico, MO) 22, no. 45 (February 24, 1881): 2.

128. *Pascagoula Democrat-Star* (Pascagoula, MS) 31, no. 12 (June 10, 1881): 1.

129. *Dallas Daily Herald* 26, no. 161 (June 3, 1881): 2.

130. Ibid., no. 240 (September 3, 1881): 2.

131. *Abbeville Press and Banner* (Abbeville, SC) 27, no. 4 (June 28, 1882): 3.

132. *Milan Exchange* (Milan, TN) 10, no. 13 (May 26, 1883): 5.

133. *Breeder's Gazette* (Chicago, IL) 6, no. 13–14 (September 25, 1884): 456.

134. David. L. Wheeler, "The Blizzard of 1886 and Its Effect on the Range Cattle Industry in the Southern Plains," *Southwestern Historical Quarterly* 94, no. 3 (January 1991): 431–432. Wheeler claimed that the impact of the 1886 blizzard resulted in several changes to cattle operations beyond the removal of barbed wire drift fences,

including the use of barbed wire for "farm and field fences" to contain stock rather than to block drifting animals.

135. Charles R. Geisst, *Monopolies in America: Empire Builders and Their Enemies from Jay Gould to Bill Gates* (New York: Oxford University Press, 2000), 42, 15–16.

136. *Weekly Chillicothe Crisis* (Chillicothe, Livingston, County, MO) 1, no. 26 (February 21, 1878): 4.

137. *Mower County Transcript* (Austin, Mower County, MN) 11, no. 38 (December 19, 1878): 1.

138. *Weekly Chillicothe Crisis* (Chillicothe, Livingston County, MO) 2, no. 28 (March 6, 1879): 4.

139. *Canton Advocate* (Canton, Dakota Territory) 5, no. 39 (January 13, 1881): 2. The *Advocate*'s masthead motto indicates a striking mission: "Evil News Rides Fast, While Good News Baits!"

140. Ibid.

141. *Omaha Daily Bee* (Omaha, NE) 10, no. 198 (February 8, 1881): 3. The Farmers' Alliance was founded in 1877 in Texas and spread throughout the South, North, and West, consolidating agrarian organizations into the National Farmers Alliance and Industrial Union with a membership of one million by 1890. See Lawrence Goodwyn's classic study, *Democratic Promise: The Populist Moment in America* (Oxford: Oxford University Press, 1976).

142. *Omaha Daily Bee* (Omaha, NE) 10, no. 198 (February 8, 1881): 3.

143. *Daily Globe* (St. Paul, MN) 4, no. 94 (April 4, 1881): 2.

144. *River Press* (Fort Benton, Montana Territory) 1, no. 26 (April 20, 1881): 1.

145. *Jamestown Alert* (Jamestown, Stutsman County, Dakota Territory) 4, no. 5 (August 26, 1881): 2.

146. *Dallas Daily Herald* 28, no. 279 (October 19, 1881): 5.

147. *Sacramento Daily Record-Union* (Sacramento, CA), *Union* 56, no. 9553 (November 12, 1881): 3; *Record* 24, no. 4557 (November 12, 1881): 3.

148. McCallum and McCallum, *The Wire That Fenced the West*, 106–107.

149. Sherman Antitrust Act, Sec. 2, 51st Cong., 1st sess., 1890, http://www.stern.nyu.edu/networks/ShermanClaytonFTC_Acts.pdf.

150. See discussion of the federal government's initially ineffective use of the Sherman Act to break trusts in Geisst, *Monopolies in America*, 57–62.

151. *Weekly Kansas Chief* (Troy, KS) 21, no. 25 (December 6, 1877): 3.

152. *Ottawa Free Trader* (Ottawa, IL) 39, no. 45 (June 7, 1879): 2.

153. See the *Fayetteville Observer* (Fayetteville, TN) 26, no. 17 (June 19, 1879): 1; *True Northerner* (Paw Paw, MI) 25, no. 16 (June 20, 1879): 6; and *Holt County Sentinel* (Oregon, MO) 15, no. 2 (June 27, 1879): 2.

154. *Daily Globe* (St. Paul, MN) 5, no. 94 (April 4, 1883): 3.

155. *Rock Island Argus* (Rock Island, IL) 31, no. 283 (September 19, 1883): 3. See, for example, *Times and Democrat* (Orangeburg, SC) 12, no. 30 (September 20, 1883): 1.

156. *Chicago Daily Tribune* (Chicago, IL) 39 (March 2, 1879): 8.

157. *News and Citizen* (Morrisville and Hyde Park, VT) 10, no. 28 (September 28, 1882): 3.

158. *Evening Star* (Washington, DC) 62, no. 9500 (October 3, 1883): 7.

159. *Dodge City Times* (Dodge City, KS) 6, no. 3 (May 31, 1879): 8.

160. *Butler Weekly Times* (Butler, MO) 5, no. 44 (October 3, 1883): 7.

161. *Omaha Daily Bee* (Omaha, NE) 13, no. 122 (November 7, 1883): 4.

162. *Shenandoah Herald* (Woodstock, VA) 63, no. 21 (March 14, 1883): 4.

163. *Middlebury Register* (Middlebury, VT) 44, no. 14 (July 4, 1879): 3.

164. *Ottawa Free Trader* (Ottawa, IL) 45, no. 1 (July 28, 1883): 7.

165. *Emporia News* (Emporia, KS) 23, no. 7 (February 13, 1880): 4.

166. *Edgefield Advertiser* (Edgefield, SC) 47, no. 39 (August 31, 1882): 2.

167. *Sedalia Weekly Bazoo* (Sedalia, MO) 14, no. 52 (May 15, 1883): 8.

168. *Helena Weekly Herald* (Helena, Montana Territory) 17, no. 2 (November 30, 1882): 2.

169. Breckenridge News (Cloverport, KY) 8, no. 16 (November 7, 1883): 2.

170. *Butler Citizen* (Butler, PA) 21, no. 2 (November 21, 1883): 1.

171. *Emporia Weekly News* (Emporia, KS) 26, no. 31 (August 2, 1883): 1.

Chapter 2

1. See Hayter, "Barbed Wire Fencing," 193–194; and McCallum and McCallum, *The Wire That Fenced the West*, 66–67. For the wording of two Texas laws that required the modification of barbed wire fences, see H. P. N. Gammel, comp., *The Laws of Texas 1822–1897* (Austin: Gammel, 1898), 107. Barbed fence laws appeared in Iowa in 1876 (amended 1878), Minnesota in 1878, Wisconsin in 1878 (amended 1880), and Washington Territory in 1879. Restrictive bills were also proposed in the 1880–1881 Maine legislative session and were referred to the Judiciary Committee, which did not recommend legislation after its investigation. See C. M. Vandeburg, "Barbed Wire," BBHC.

2. The three statements examined include *Barb Fencing. Statement of the Washburn & Moen Manufacturing Co., before the Committee on Agriculture of the General Assembly of Connecticut, Thursday, February 9, 1880* (Worcester, MA: Press of Snow, Woodman, 1880); *A Manual of the Fence: Comprising the Fence Complaint of the Past, and the Fence Remedy Assured in Steel Barb Fencing* (Worcester, MA: Washburn & Moen, 1881); and *Barb Wire Fencing: A Statement of the Washburn & Moen Manufacturing Co. to the Legislature of New Hampshire, July 12, 1881* (Manchester, NH: John B. Clarke, 1881).

3. *Hartford Daily Courant* (Hartford, CT) 44, no. 50 (February 27, 188): 2.

4. Ibid.

5. "Part 2: Barb Fencing. Statement of the Washburn & Moen Manufacturing Co., before the Committee on Agriculture of the General Assembly of Connecticut, Thursday, February 19, 1880," in *A Manual of the Fence*, 3.

6. Ibid., 3–20.

7. Ibid., 25.

8. Ibid.

9. Ibid., 27.

10. *Hartford Daily Courant* (Hartford, CT) 44, no. 50 (February 27, 1880): 2.

11. *Vermont Watchman & State Journal* (Montpelier, VT) 76–3867, no. 7 (November 24, 1880): 1.

12. Ibid.

13. Ibid. The second House bill (HB 259) required that for any barbed wire fencing constructed along a highway or between properties, all adjoining landowners must provide written consent.

14. *Journal of the House of Representatives of the State of Vermont: Biennial Session, 1880* (Montpelier, VT: Freeman Steam Printing House and Bindery, 1881), 140–141, 169, 341, 365.

15. *Journal of the Senate of the State of Vermont* (Montpelier, VT: Freeman Steam Printing House and Bindery, 1881), 89, 319.

16. *Vermont Watchman & State Journal* (Montpelier, VT) 76–3867, no. 7 (November 24, 1880): 1.

17. Ibid.

18. Ibid.

19. Ibid.

20. "Part 3: The Barb Fence Hearing in Vermont; Is It a Cruel Fence? Barb Fence Hearing before the Joint Committee of the Vermont General Assembly," in *A Manual of the Fence*, 7.

21. *Glidden Patent Two Point and Four Point Barb Wire Fencing: Cheap, Staunch, Indispensable* (Worcester, MA: Washburn & Moen, n.d.), 12, BBHC.

22. "Part 3: The Barb Fence Hearing," 7.

23. Ibid., 8.

24. Ibid.

25. In his account, Vandeburg gave only a passing reference to the society's role in supporting the restrictive measure, so we can only speculate that there was enough pressure to force the legislature to take action twice. Vandeburg, "Barbed Wire."

26. *Barb Wire Fencing*, 13.

27. Ibid., 14.

28. Vandeburg, "Barbed Wire."

29. *Barb Wire Fencing*, 14.

30. "The Barbed Wire Cruelty—a Nuisance That Needs Abating," *Our Dumb Animals* 14, no. 7 (December 1881): 55. The article was reprinted from the *Lancaster* (PA) *Inquirer*. The Humane Society of the United States still advocates that "the pasture should be bordered by safe, sturdy fencing. . . . Barbed wire is not an acceptable fencing material—it has been the cause of many serious injuries." See "Horse Care Guidelines," accessed March 8, 2017, http://www.humanesociety.org/animals/horses/tips/horse_care_guidelines.html.

31. "The Barbed Wire Cruelty," 55. Not all subscribers were as indignant. In a letter to the editor the following April, a subscriber responded to the article: "Is barb wire

cruel to animals? In every case, after all had been said that could be said, the verdict has been, No. The most experienced farmers say that there are fewer and less serious accidents from barb wire than from the old styles of fences." *Our Dumb Animals* 14, no. 11 (April 1882): 83.

32. ACME Veterinary Wire Cut Remedy advertisement, *Sears, Roebuck, & Co., Cheapest Supply House on Earth*, catalogue no. 109 (Chicago, IL: Sears, Roebuck, 1899), 26.

33. A photo of the original image can be seen at the Ellwood House Museum, DeKalb, Illinois (hereafter EHM). The museum has an interesting display of bottles of various brands of barbed wire liniment manufactured in the late nineteenth century.

34. *Haller Catalogue, Health Guide, & Almanac, 1908* (Blair, NE: Haller Proprietary, 1908), Kansas Barbed Wire Museum, La Crosse, KS (hereafter KBWM).

35. Scarless Liniment, Scarless Remedy Company, Winterset, IA, KBWM.

36. See, for example, Glover's Barbed Wire Liniment: For Man and Beast, Cuts, Bruises, Running Sores, Internal Use, Tape Worms, Scours, and Gleet–Jack Glover, Ft. Gibson, Cherokee Nation, KBWM; Perrigo's Barb Wire Liniment For Man or Beast, L. Perrigo Co., Allegan, MI, KBWM; Dr. J.W. Pinkerton Barbed Wire King for Wire Cuts, Old Sores, Sore Shoulder, Sore Teats, Scratches, Grease Heal, Any and All Abrasions for the Skin. Good for Man or Beast, KBWM.

37. *The Fence Problem in the United States, as Related to General Husbandry and Sheep Raising: Facts and Statistics from Authoritative Sources with a View of Fence Laws and Customs, Illustrated* (Worcester, MA: Washburn & Moen, 1882), 8.

38. *Glidden Patent Steel Barb Wire Fencing*, (Worcester, MA: Washburn & Moen, n.d.), BBHC.

39. "The Principles of Steel Barb Fencing," in *Steel Barb Fencing*, 17–18.

40. *De Kalb County Manufacturer*, 24.

41. *Steel Barb Fencing*, 27–28.

42. Ibid., 7.

43. *Glidden Barb-Fence Journal* (DeKalb, IL: I. L. Ellwood, 1886), 16.

44. Ibid. (1884), 2, 14.

45. Ibid., 12.

46. Ibid. (1886), 13.

47. "Barbed Wire for Fencing," *Chicago Daily Tribune* (Chicago, IL) 41 (December 12, 1881): 6.

48. *Glidden Barb-Fence Journal* (1886), 7.

49. Editorial, *Chicago Daily Tribune* (Chicago, IL) 40 (April 6, 1881): 4, cited in *The Fence Problem*, 45–46. Ironically, during the previous summer the City Council of Chicago attempted to pass a city ordinance that would forbid the use of barbed wire. After the 27–3 vote to ban barbed wire within the city limits, the mayor announced without explanation that he would veto the ordinance. See "The Council," *Chicago Daily Tribune* (Chicago, IL) 40 (June 29, 1880): 8.

50. McCallum and McCallum, *The Wire That Fenced the West*, 232–238.

51. William D. Hunt, Improvement in Fences, US Patent 67,117, no filing date, issued July 23, 1867; Michael Kelly, Improvement in Fences, US Patent 74,379, no

filing date, issued February 11, 1868; Rose, Improvement in Wire Fences, US Patent 138,763; Jacob Haish, Improvement in Barbed Fences, US Patent 147,634, filed December 27, 1873, issued February 17, 1874; Joseph F. Glidden, Improvement in Wire-Fences, US Patent 157,124, filed October 27, 1873, issued November 24, 1874; Isaac L. Ellwood, improvement in Barbed Fences, US Patent 147,756, filed January 7, 1874, issued February 24, 1874; Charles Kennedy, Improvement in Barbed Fences, US Patent 153,965, filed June 2, 1874, issued August 11, 1874; Luther Merrill and John C. Merrill, Improvement in Wire Fences, US Patent 155,538, filed August 28, 1874, issued September 29, 1874; Aaron Nellis, Improvement in Barbed Fences, US Patent 187,723, filed January 18, 1877, issued February 27, 1877; Henry W. Putnam, Improvement in Barbed Fence-Wire, US Patent 187,172, filed January 20, 1876, issued February 6, 1877; John McNeill, Improvement in Barbed Fence-Wires, US Patent 189,122, filed July 18, 1876, issued April 3, 1877.

52. Parker Wineman, Improvement in Barbed Fence-Wires, US Patent 176,725, filed March 10, 1876, issued April 25, 1876.

53. Spencer H. St. John, Improvement in Barbs for Fence-Wire, US Patent 199,330, filed December 13, 1877, issued January 15, 1878; Montraville P. Mighell, Improvement in Barbed Fence-Wires, US Patent 199,924, filed July 25, 1877, issued February 5, 1878; William Warden, Improvement in Barbed Fences, US Patent 199,947, filed September 27, 1877, issued February 5, 1878.

54. In order granted: Charles F. Washburn, Improvement in Barbed Fence-Wires, US Patent 200,494, filed November 17, 1876, issued February 19, 1878; Lewis T. Larson Wing, Improvement in Fence-Barb, US Patent 200,783, filed July 9, 1877, issued February 26, 1878; Thomas H. Dodge, Improvement in Barbed Fence-Wires, US Patent 201,507, filed March 6, 1878, issued March 19, 1878; and John W. Nadelhoffer, Improvement in Barbs for Wire Fences, US Patent 201,889, filed December 13, 1877, issued April 2, 1878.

55. Lucien B. Smith, Wire Fence, US Patent 66,182, no filing date, issued June 25, 1867.

56. Kelly, Improvement in Fences, US Patent 74,379.

57. Edward M. Crandal, Improvement in Indicators for Barbed-Wire Fences, US Patent 220,912, filed August 23, 1879, issued October 28, 1879.

58. Hiram B. Scutt, Warning-Plate for Barbed-Wire Fences, US Patent 224,482, filed April 7, 1879, issued February 10, 1880. Arthur G. Hulbert was issued a similar patent in 1884 that consisted of wood blocks twisted between two wires with "a barb of metal set transversely through its [each] center." See Wire Fence, US Patent 296,835, filed February 24, 1883, issued April 15, 1884.

59. John J. La Fleur, Tag for Barbed-Wire Fences, US Patent 249,777, filed July 5, 1881, issued November 22, 1881.

60. Pounds of barbed wire made and sold in 1874, 10,000; 1876, 2.84 million; 1878, 19.655 million; and 1880, 41 million. See *Barb Wire Fencing*, 10.

61. Thomas G. Orwig, Improvement in Barbed Fence-Wires, US Patent 201,890, filed August 20, 1877, issued April 2, 1878.

62. Nadelhoffer, Improvement in Barbs for Wire Fences, US Patent 201,899.

63. William W. Butler, Fence-Barb, US Patent 248,999, filed August 27, 1881, issued November 1, 1881.

64. John M. May, Barbed-Wire Fence, US Patent 264,728, filed April 30, 1879, issued September 19, 1882; George M. Beerbower, Barbed Fence, US Patent 327,755, filed October 20, 1884, issued October 6, 1885; Edward W. Robinson, Barb for Wire Fences, US Patent 421,055, filed March 8, 1889, issued February 11, 1890; Constant S. Trevitt and Solomon F. Mouck, Barbed Fence, US Patent 404,879, filed January 28, 1889, issued June 11, 1889.

65. Jay Bostwick Fisher, Improvement in Fences, US Patent 203,536, filed December 5, 1877, issued May 14, 1878. The complexity of Fisher's patent ruled out the possibility of mass production.

66. Hugh Gearty, Barb for Wire Fences, US Patent 472,044, filed December 26, 1891, issued April 5, 1892.

67. Fitz Randolph Longstreet, Wire Fence, US Patent 711,574, filed December 7, 1901, issued October 21, 1902. Although slightly beyond the nineteenth century, Longstreet's patent represented the zenith of safety innovations in barbed fencing.

68. There were of course exceptions to the chronological grouping of patents into the ones deterring contact through pricking and the ones avoiding the language of pain. Among the patents noted in Henry and Frances McCallum's "Early Varied" category, several patents were granted that included specific references to the prevention of serious injury to stock, four patents in 1877 alone. William Reynolds claimed that his zinc barbs would not rust and therefore create "permanent injury." John McNeill stated his beveled pointed barbs would "possess a sufficiently offensive character for cattle, and . . . avoid the danger of injuring them by penetrating too far." Charles Parker's patent claimed its "elastic fence" would "not so seriously injure stock that might run against it." Lyman Judson's patent possessed barbs "loosely clasped" on a metal strip "that will yield sufficiently to not tear the flesh of an animal." And finally one year later in 1878, Jay Fisher declared his angular or blunt-ended barbs would not "readily penetrate the skin or flesh of the animals." See William Reynolds, Improvement in Barbed Fence-Wire, US Patent 187,049, filed September 23, 1876, issued February 6, 1877; McNeill, Improvement in Barbed Fence-Wires, US Patent 189,122; Charles Parker, Improvement in Barbed Fences, US Patent 194,260, filed April 10, 1877, issued August 4, 1877; Lyman Judson, Improvement in Barbed Fences, US Patent 191,348, filed March 24, 1877, issued May 29, 1877; and Fisher, Improvement in Fences, US Patent 203,536.

Chapter 3

1. Netz, *Barbed Wire*, 1.

2. Richard Ohmann, *Selling Culture: Magazines, Markets, and Class at the Turn of the Century* (New York: Verso, 1996), 11–29.

3. Pamela Walker Laird, *Advertising Progress: American Business and the Rise of Consumer Marketing* (Baltimore: Johns Hopkins University Press, 1998), 462–463.

4. Robert Jay, *The Trade Card in Nineteenth-Century America* (Columbia: University of Missouri Press, 1987), 34.

5. Cary Nelson and Michael Chasar, "American Advertising: A Poem for Every Product," in *The Oxford History of Popular Print Culture*, ed. Christine Bold, 6 vols. (Oxford: Oxford University Press, 2012), 6:134.

6. Aleksandr Herzen, quoted in Jackson Lears, *Fables of Abundance: A Cultural History of Advertising in America* (New York: Basic Books, 1994), 54.

7. Laird, *Advertising Progress*, 44.

8. Nelson and Chasar, "American Advertising," 135.

9. Washburn & Moen Manufacturing Co. and I. L. Ellwood & Co., *Barb Fence: Its Utility, Efficiency and Economy: A Book for the Farmer, the Gardener and the Country Gentleman* (Worcester: Goddard, 1876), 3, 4–5, 5.

10. Ibid., 7–8. An 1882 Glidden "Pocket Compendium" leaves out the country gentleman and proclaims that its fence is "For the Farmer, Gardener, Stockman, Railroad Co.'s," indicating a shifting focus from the East to the West, EHM.

11. *Barb Fence*, 1876, 11.

12. Ibid., 11, 37.

13. Ibid., 13–28.

14. Ibid., 19.

15. Ibid. A letter from the firm of Kirkham & Macgregor further undermines the spontaneous nature of the testimonies, for the company intends to "procure a testimonial from every farmer to whom we have sold your wire. . . . We will only send you the certificates of some six or eight influential farmers who are using this wire. . . . We will forward [certificates] as soon as we can see the farmers." The firms of Higley & Bro. and Cunningham & Anderson also forwarded letters from their clients as early as 1875. Ibid., 13–14, 16, 18.

16. Ibid., 14.

17. Ibid., 25.

18. Ibid., 23.

19. Ibid., 25.

20. Ibid., 20.

21. Ibid., 27.

22. Ibid., 39.

23. *Barb Fence: Its Utility, Efficiency, and Economy: A Book for the Farmer, the Gardener, and the Country Gentleman* (Worcester: Noyes, Snow, 1878), 39–40.

24. Ibid., 58.

25. *The Fence: A Compilation of Facts, Figures and Opinions from National and State Agricultural Reports, Agricultural Journals, and the Public Press, for the Past Sixty Years, (Presented in Their Original Form,) Showing the Waste and Need of Fencing Material, the Cost of Fences, Their Waste and Wear in the United States, Relation to Other Farm Property, etc., etc. and a Compilation of Statutory and Judicial Facts and Decisions in the State and Federal Courts, Relating to Railway Fences and Fence Laws, with the Early History of Enclosures* (Worcester, MA: Press of Noyes, Snow, 1879), 8.

26. Ibid., 41.

27. *Steel Barb Fencing*, 8, 10, 12, 20.

28. *The Fence Problem*, 5.

29. Jacob Haish's *Barb Fence Regulator* 2 (1877?), and vol. 1 (1876?); and Isaac Ellwood's *Glidden Barb-Fence Journal* 1 (1880?).

30. Haish, *Reminiscent Chapter*, 14.

31. Because Haish's records were damaged in a flood and subsequently destroyed, these numbers cannot be verified. Circulation figures for the *Glidden Barb-Fence Journal* could not be found in the American Steel & Wire Company records located in the Baker Library of Harvard University.

32. Haish, *Reminiscent Chapter*, 14. As noted earlier, Haish was not the first to produce an almanac to sell his product, just the first to do so in the barbed wire business.

33. Robert Huhn Jones, "Three Days of Violence: The Regulators of the Rock River Valley," *Journal of the Illinois State Historical Society* 59 (1966): 131–142. Haish arrived in Illinois in 1845, four years after the vigilante movement, which was no doubt still fresh in the public's mind. Haish moved to DeKalb eight years later. See also a brief history in "Barbed Wire Inventors: Jacob Haish," Antique Barbed Wire Society, accessed March 12, 2017, http://www.antiquebarbedwiresociety.com/inventors_haish.html.

34. *Barb Fence Regulator* 2 (1877?): 1.

35. Ibid. 6 (1882 [1883?]): 5.

36. Ibid. 3 (n.d.): 4. Haish relied on the American stereotypes of German immigrants as industrious, progressive, and commonsensical. For discussions of German immigrants and how they were perceived in a multicultural nation, see Frederick Luebke, *Ethnicity on the Great Plains* (Lincoln: University of Nebraska Press, 1980); and Terry G. Jordan, *German Seed in Texas Soil: Immigrant Farmers in Nineteenth-Century Texas* (Austin: University of Texas Press, 1966). The Pew Research Center reports that in 1870 Ireland (1.8 million) and Germany (1.7 million) were the top two foreign populations in the United States, with Germans the largest group of immigrants in the midwestern states, the early center of the barbed wire business. By 1880 Germans had replaced the Irish as the largest group of immigrants, and their share grew in numbers over the rest of the nineteenth century. See "From Ireland to Germany to Italy to Mexico: How America's Sources of Immigrants Has Changed in the States, 1850–2013," Pew Research Center, September 28, 2015, http://www.pewhispanic.org/2015/09/28/from-ireland-to-germany-to-italy-to-mexico-how-americas-source-of-immigrants-has-changed-in-the-states-1850-to-2013/.

37. Nelson and Chasar, "American Advertising," 135.

38. *Barb Fence Regulator* 4 (1881): 3.

39. Ibid. 6 (1882 [1883?]): 5; see also vols. 3, 4.

40. Ibid. 2 (1877?): 7.

41. Ibid., 3.

42. Ibid., 4.

43. According to the 1981 edition of *Compact Edition of the Oxford English Dictionary*, "bull-dose, doze" is "a severe dose of flogging; to coerce by violence, intimidate. . . . If a negro is invited to join it [a society called 'The Stop'], and refuses, he is taken to the woods and whipped. This whipping is called a 'bull doze,' or doze fit for a bull. The application of the bull-doze was for the purpose of making Tilden voters" (vol. A–O, 1168).

44. *Barb Fence Regulator* 2 (1877?): 5.

45. Ibid. 4 (1881): 1.

46. Ibid.

47. Ibid., 7.

48. Ibid. 6 (1882 [1883?]): 1, 2, 10.

49. I. L. Ellwood, letter to Washburn & Moen Co., received by Washburn August 23, 1881, BBHC.

50. *Glidden Barb-Fence Journal* 2 (1881): 8.

51. Ibid., 5 (1884): 3, 10.

52. Ibid. 2 (1881): 4.

53. Ibid., 16.

54. Lears, *Fables of Abundance*, 145.

55. *Barb Fence Regulator* 4 (1881): 8.

56. Lears, *Fables of Abundance*, 144.

57. The text on the side of the train car in another version of the image promises violent opposition: "Der vatch-togs shoot all / Vat look like opposers." *Barb Fence Regulator* 4 (1881): 8.

58. Laird, *Advertising Progress*, 3, 130–132.

59. *Barb Fence Regulator* 11 (1887): 16.

60. Laird, *Advertising Progress*, 132; for more about factories, see 127–128.

61. *Steel Barb Fencing*, 8.

62. *Barb Fence,* 1876, 11, 36–38.

63. *Glidden Barb-Fence Journal* 6 (1885) 15.

64. During the Connecticut legislature's discussion of restricting barbed wire usage, a Washburn & Moen representative claimed that "wire fencing is sold from Egypt to Oregon." *Connecticut Farmer* 2, no. 2 (March 6, 1880): 5.

65. In the wake of the Civil War, Anglo westward expansion led to conflict with the Great Plains Native cultures. The raids and massacres of the late 1860s through the late 1870s resulted in calls for the eradication of Indian cultures. Federal policy focused on incarcerating Indians (eventually regardless of previous land cession treaties) on reservations where religious groups were tasked with assimilating them by force, if necessary, into mainstream society. The deterioration of reservation conditions and broken treaty promises resulted in many Native Americans raiding off reservations well into the 1880s. But the near extermination of buffalo, the onslaught of disease, and the army's relentless pursuit weakened Indians' ability to survive off the reservations. See Richard White, *"It's Your Misfortune and None of My Own": A New History of the American West* (Norman: University of Oklahoma Press, 1991), 103–109.

66. McCallum and McCallum, *The Wire That Fenced the West*, 117.

67. *Glidden Barb-Fence Journal* 8 (1887): 2.

68. See White, *"It's Your Misfortune,"* 104–108, for an account of what he calls "completing the conquest."

69. *Iron Age* 50 (December 1, 1892): 1071.

70. Krell, *The Devil's Rope*, 39. For correspondence on the selection of Waukegan model, see letter dated March 2, 1936, sent by "Dodson" to "Van" from Torreon, Coahuila, Mexico, American Steel and Wire Company Collection, BBHC. Waukegan wire was first produced in 1890 in the era of the waning Indian Wars, Reservation Policy, and the last nineteenth-century revitalization movement on the Plains (Ghost Dance).

71. "Why Barb Fencing Is Better Than Any Other," *Glidden Barb-Fence Journal* 6, no. 6 (1885): 15.

72. Joseph Boskin, *Sambo: Rise & Demise of an American Jester* (Oxford: Oxford University Press, 1986), 16.

73. *Barb Fence Regulator* 3 (1879): 4.

74. *Glidden Barb-Fence Journal* 2 (1881): 4, continues on this line of racist provocation: "De fac is Sambo dat dose 'ere mellons have dun gone and got b'hin wun of dem Glidden Barb Fences, an' spite 'f fate dis indiwidual could-n't appropr'ate de fust one. Dod rot dat fence no how! It's gettin so niggers dont have no mo show'n white pussons."

75. Lears, *Fables of Abundance*, 87.

76. *Barb Fence Regulator* 4 (1881): 1.

77. "Uncle Remus on Education," in ibid., 6. For more on African Americans in late nineteenth-century advertising, see Ohmann, *Selling Culture*, 255–266.

78. Jay, *The Trade Card*, 68, 70.

79. Jane Marcellus, "Nervous Women and Noble Savages: The Romanticized 'Other' in Nineteenth-Century US Patent Medicine Advertising," *Journal of Popular Culture* 41, no. 5 (2008): 791, 803.

80. Arthur G. Miller, ed., *In the Eye of the Beholder: Contemporary Issues in Stereotyping* (New York: Praeger, 1982), 95.

81. For a fascinating account of Wilde's visit to the American West, see Jan Wellington, "Oscar Wilde's West," *Literary Traveler*, August 2007. American publications alternated between laudatory and scathing remarks about Wilde's tours of the United States. The *Richmond Democrat* (Missouri) noted in 1882 that "down in Texas they call Oscar Wilde 'colonel.' When he gets over into Alabama they will call him 'jedge' and in Kentucky he will be known as 'gineral.' In Lathrop he will be known buy his right name 'damphool.'" *Richmond Democrat* 10, no. 29 (July 20, 1882): 2. A year later the newspaper came back to Wilde in an odd story that casts Wilde as heterosexual: "Does any one recognize this picture? He used to be one of Richmond's favorite dudes, but Sunday evening he went to see his best girl, and as ill luck would have it, he and Frances sat down on one of those new-fangled sofas—the ones that break in two at ten o'clock, you know—and his dejected appearance is attribut-

able to that fact alone. The thing went off at the moment indicated, and as Frances was wafted to her little couch up stairs, Oscar Wilde moved unserenely out the front door and scaled the barb-wire fence in a manner that did honor to-the patent sofa." Ibid. 11, no. 29 (July 19, 1883): 3.

82. *Glidden Barb-Fence Journal* 6 (1885): 7.

83. Ibid.

84. Ibid. 2 (1881): 12.

85. Ibid., 7.

86. Ibid. 5 (1884): 3.

87. Ibid., 7. This particular turn of phrase deftly combined sexism and racism to demean women of color.

88. Ibid., 8, 12.

89. Ibid. 6 (1885): 3.

90. *Barb Fence Regulator* 11 (1887): 6.

91. "Ancient Ancestry," ibid. 2 (1877?): 8; "To the Trade," ibid. 4 (1881): 1.

92. Ibid. 6 (1882 [1883]?): 4.

93. Ibid., 12.

94. Ibid. 11 (1887): 4.

95. Quoted in Roland Marchand, *Advertising the American Dream: Making Way for Modernity, 1920–1940* (Berkeley: University of California Press, 1985), 265.

96. This image was also printed in the 1887 *Barb Fence Regulator*. Krell discusses the "Edenesque allusions" of the image in *The Devil's Rope*, 14–15.

97. *Glidden Barb Fence Journal* 2 (1881): 9.

98. Ibid. 2 (1881): 10. Haish also refers to Samson's use of an ass's jawbone, with emphasis on the ass: "For the same weapon has been in constant use by the Monopolists for the past three years and more and yet none of the Philistines have fallen. . . . We have concluded that the educated thought and mature judgment of the nineteenth century . . . ignores such worn-out and useless weapons to decide questions of righteousness and justice." See *Barb Fence Regulator* 4 (1881): 2.

99. *Barb Fence Regulator* 11 (1887): 1.

100. Ibid. The Manz Company, located in Chicago, produced images for both Haish and Ellwood.

101. *Glidden Barb-Fence Journal* (1881, 1884, 1885, 1886, 1887, 1888): covers. The order created by the barbed wire fences in this scene echoes the order found in late nineteenth-century farmhouses. See Sally Ann McMurry, *Families and Farmhouses in Nineteenth-Century America: Vernacular Design and Social Change* (Oxford: Oxford University Press, 1988); Fred W. Peterson, "Vernacular Building and Victorian Architecture: Midwestern American Farm Homes," *Journal of Interdisciplinary History* 12 (Winter 1982): 409–427.

102. Paul G. Bourcier, "'In Excellent Order': The Gentleman Farmer Views His Fences, 1790–1860," *Agricultural History* 58, no. 4 (October 1984): 546–547.

103. *Transactions of the Iowa State Horticultural Society for 1888: Twenty-Third Annual Session Held at Des Moines, January 15, 16, 17, 1889. Also Proceedings of the West-*

ern, Eastern and Northern Societies, with Summary of Other Local Societies for the Year 1888, ed. Geo. Van Houton (Des Moines: Geo. E. Roberts, State Printer, 1889), 23:309.

104. Bourcier, "In Excellent Order," 555, 560.

105. *Glidden Barb-Fence Journal* 8 (1887): 1.

106. Ibid. 5 (1884): 5.

107. Ibid. 8 (1887): 7.

108. "Barbed Fence Wire," circular, J. W. Gates & Co., Wire Merchants, 1882, BBHC.

109. *De Kalb County Manufacturer*, 14, 16.

110. The back cover of the 1887 *Regulator* similarly contrasts crude and uncertain wilderness with prosperous and civilized domesticity.

111. Alphonso Dabb, Improvement in Pickets for Fences and Walls, US Patent 63,482, no filing date, issued April 2, 1867.

112. Royal C. Eaton, Barbed Picket Fence, US Patent 304,630, filed March 1, 1884, issued September 2, 1884; James Locher, Picket and Barbed Fence, US Patent 388,780, filed November 21, 1887, issued August 28, 1888; and Henry W. Volck, Picket Fence, US Patent 431,759, filed March 10, 1890, issued July 8, 1890.

113. James Emerson's patent for a metal picket for fences, although not barbed, clearly replicated the ideal domestic picket fence using metal pickets with serrated edges: Metal Picket for Fences, US Patent 421,685, filed December 13, 1889, issued February 18, 1890.

114. Barbara Welter, "The Cult of True Womanhood, 1820–1860," *American Quarterly* 18 (1966): 152.

115. See *Barb Fence Regulator* 4 (1881): 4, which uses the word "soda" instead of the original "lager."

116. See, for example, two images in *Barb Fence Regulator* 2 (1877), masthead, 8; 4 (1881): 2, 6–7; 6 (1882 [1883?]): masthead, 3; 11 (1887): cover, 1, 11; and 14 (n.d.): cover, 1.

117. Ibid. 11 (1887): 8.

118. In 1901 Ellwood Steel Fences produced a trade card that showed the various kinds of fences they manufactured in yet another domestic bucolic scene. However, most of the wire fence used around the house yard and horse and hog pastures consisted mainly of woven wire, a harmless steel netting. At the top of the fences hang two strands of barbed wire much less vicious than the four to six strands used previously. Ellwood Steel Fences, trade card, 1901, EHM.

Part 2

1. Appadurai, *Social Life of Things*, 3.

2. Bill Brown, *A Sense of Things: The Object Matter of American Literature* (Chicago: University of Chicago Press, 2003), 4.

3. Mark Twain, *A Connecticut Yankee in King Arthur's Court* (New York: Modern Library, 2001), 211.

4. Ibid., 405.

5. Justin Kaplan points out that while there is no mention of barbed wire in this battle scene of the published novel, when Twain gave an 1886 public reading of the

first three chapters, he was planning to arm the fence with barbs as well as electricity. Kaplan, *Mr. Clemens and Mark Twain: A Biography* (New York: Simon & Schuster, 2006), 295.

6. Netz, *Barbed Wire*, 102.

7. Twain, *Connecticut Yankee*, 444–447.

8. Erich Maria Remarque, *All Quiet on the Western Front*, trans. A. W. Wheen (New York: Fawcett Crest, 1975), 112–113.

9. See the biographical sketch of Haish at "Barbed Wire Inventors: Jacob Haish," Antique Barbed Wire Society, accessed March 12, 2017, http://www.antiquebarbedwiresociety.com/inventors_haish.html.

10. Luis Alberto Urrea, *The Hummingbird's Daughter* (Boston: Little, Brown, 2005), 24.

11. Bruce Nauman, *Setting a Good Corner (Allegory & Metaphor)*, dir. Bruce Nauman (Galisteo, NM: Private Production, 2000).

12. For the interview with Nauman about the film, see "Setting a Good Corner: Bruce Nauman," ART21, November 2011, http://www.ART21.org/read/bruce-nauman-setting-a-good-corner.

13. Elliot Lincoln, *Rhymes of a Homesteader* (Boston: Houghton Mifflin, 1920), 75.

14. Amy Clampitt, *The Kingfisher* (New York: Knopf, 1982), 72–73.

15. Quoted in Mary Jo Salter, introduction to *The Collected Poems of Amy Clampitt* (New York: Knopf, 1999); also see Amy Clampitt Fund, 1997, http://www.amyclampitt.org/biography/salter.html.

16. Krell, *The Devil's Rope*, 115, 147.

17. Erwin Panofsky, *Studies in Iconology* (New York: Oxford University Press, 1939), 11.

18. Theodore Ziolkowski, *Disenchanted Images: A Literary Iconology* (Princeton, NJ: Princeton University Press, 1977), 13. Ziolkowski studies the haunted portrait and the magic mirror over the historical course of their disenchantment in this text.

19. Theodore Ziolkowski, *Varieties of Literary Thematics* (Princeton, NJ: Princeton University Press, 1983), ix. In this work Ziolkowski takes the literary measure of teeth, the mystic carbuncle, and talking dogs, to mention only a few the motifs.

Chapter 4

1. *Austin Weekly Statesman* (Austin, TX) 13, no. 14 (December 6, 1883): 1.

2. Davis wrote for newspapers in Tyler, Galveston, and Houston, Texas, and for the *New Orleans Picayune*. She was well known for her poetry, novels about plantation life in Louisiana, and essays.

3. Mollie E. Moore Davis, *The Wire Cutters* (1899; repr., College Station: Texas A&M University Press, 1997), 93, 94–95, 101–102.

4. Ibid., 108–109.

5. Ibid., 94–95.

6. Ibid., 133, 136.

7. Quoted in Krell, *The Devil's Rope*, 43. See *Fort Worth Daily Gazette* (Fort Worth, TX) 8, no. 303 (November 7, 1884).

8. Krell, *The Devil's Rope*, 43. See the *Galveston Daily News* (Galveston, TX) 44, no. 233 (December 13, 1885).

9. *Austin Weekly Statesman* (Austin, TX) 13, no. 4 (September 27, 1883): 8.

10. *Fort Worth Daily Gazette* (Fort Worth, TX) 7, no. 86 (March 29, 1883): 6.

11. Davis, *The Wire Cutters*, 277.

12. *Austin Weekly Statesman* (Austin, TX) 13, no. 4 (September 27, 1883): 8.

13. Frederick Jackson Turner, "The Significance of the Frontier in American History," in *The Frontier in American History*, ed. Frederick Jackson Turner (New York: Henry Holt, 1921): 1–38.

14. Richard Slotkin, *Regeneration through Violence: The Mythology of the American Frontier, 1600–1860* (Norman: University of Oklahoma Press, 1973), 3–5.

15. In his introduction to Davis's *The Wire Cutters*, Lou Halsell Rodenberger argues that "contrary to the perception of most literary historians interested in fiction of the West, it was a Texas woman writer who produced the first serious Western novel" (ix). While there is some basis for his claim, and as important as *The Wire Cutters* is for our work here, we agree with most critics that the character, setting, and plot of *The Virginian* resonate generically in ways that those of Davis's novel do not.

16. Owen Wister, *The Virginian: A Horseman of the Plains* (New York: Macmillan, 1902), 2, 13, 69, 231, 307.

17. Jennifer S. Tuttle, "Indigenous Whiteness and Wister's Invisible Indians," in *Reading* The Virginian *in the New West*, ed. Melody Graulich and Stephen Tatum (Lincoln: University of Nebraska Press, 2003), 102.

18. Wister, *The Virginian*, 97.

19. Ibid., 298.

20. Ibid., 407–408. Krista Comer argues that a panoramic view like this enables a kind of masculine mastery of the landscape. See Comer, *Landscapes of the New West: Gender and Geography in Contemporary Women's Writing* (Chapel Hill: University of North Carolina Press, 1999), 24. Our reading of this and other scenes of wilderness is related to Comer's view, although rather than a masculine mastery, we see the natural threat and subsequent overcoming of the threat as an experience of the sublime.

21. Immanuel Kant, *Critique of Judgement*, trans. James Creed Meredith (Oxford: Oxford University Press, 2007), 92.

22. Wister, *The Virginian*, 458.

23. William Handley reads this in terms of Wister's personal conflict between a national and racial imperative to marry and his own homosocial desire. He discusses the novel and its narrative structure in the context of the narrator's obvious infatuation with the Virginian and Wister's own infatuations with men of the American West. William Handley, "Wister's Omniscience and Omissions," in *Reading* The Virginian *in the New West*, ed. Melody Graulich and Stephen Tatum (Lincoln: University of Nebraska Press, 2003), 39–71.

24. Wister, *The Virginian*, 485–486, 502.

25. Zane Grey, *The Drift Fence* (New York: Grosset & Dunlap, 1933), 50–52.

26. The Pleasant Valley War (Tonto Basin War), begun in 1882, was a decade-long feud between two families, the Tewksburys and the Grahams, over water and grazing rights. Zane Grey based his 1921 novel *To the Last Man* on this feud. See Bob Boze Bell, "Classic Gunfights: To the Last Man," *True West*, 2017, http:/www.truewestmagazine.com/classic-gunfights-last-man.

27. Elmer Kelton, *Barbed Wire* (New York: Forge, 2006), 8–9, 57, 63.

28. Ibid., 63–64.

29. See Elmer Kelton, *Shotgun and Six Bits a Day* (New York: Forge, 2014), 15. In this story of rancher versus rancher, Bishop's fences not only improve his cattle (and profits), but during an extended drought these same fences serve to (at least initially) protect his claim to land and water resources within his fences. However, the pressure from drought on resources pits neighboring ranchers (Bishop and Clarence Cass) against each other, and fence cutting ensues: "'There's more than an occasional critter in that pen; count them yourself. There's somethin' past a hundred head. And they didn't drift. Finn and Hez found where they was pushed through a cut fence. There was footprints all around where that wire was pinched. Cass footprints, I'd judge.' Cass blustered. 'You got no proof of that'" (13).

30. Cameron Judd, *Devil Wire/Brazos* (New York: St. Martin's Paperbacks, 2008), 125, 41.

31. Jack Schaefer, *Shane*, ed. James C. Work (Lincoln: University of Nebraska Press, 1984), 72, 71.

32. *Shane*, dir. George Stevens (Los Angeles: Paramount Pictures, 1953).

33. Ibid.

34. Matthew Costello, "'I Didn't Expect to Find Any Fences around Here': Cultural Ambiguity and Containment in *Shane*," *Journal of American Culture* 27, no. 3 (2004): 262.

35. *Man without a Star*, dir. King Vidor (Los Angeles: Universal-International, 1955).

36. Louis L'Amour, *Kiowa Trail* (New York: Bantam Books, 2009), 42–47.

37. Maurice De Bevere (Morris) & René Goscinny, *A Lucky Luke Adventure: Barbed Wire on the Prairie*, trans. Luke Spear (Ashford, Kent, UK: Cinebook, 2007), 45.

38. Wallace Stegner, "Variations on a Theme by Crèvecoeur," in *New West Reader*, ed. Philip Connors (New York: Nation Books, 2005), 117.

39. Ibid., 118, 119.

40. Gretel Ehrlich, "The Solace of Open Spaces," in *New West Reader*, ed. Philip Connors (New York: Nation Books, 2005), 98–99.

41. *HUD*, dir. Martin Ritt (Los Angeles: Paramount Pictures, 1963).

Chapter 5

1. Interpretations of the New West gained ground steadily throughout the late twentieth century. An early and still relevant synthesis is found in Patricia Limerick, *The Legacy of Conquest: The Unbroken Past of the American West* (New York: W. W.

Norton, 1987). Another important book in this regard is Patricia Nelson Limerick, Clyde A. Milner II, and Charles E. Rankin, eds., *Trails: Toward a New Western History* (Lawrence: University Press of Kansas, 1991).

2. Edward Abbey, *The Monkey Wrench Gang* (New York: Harper Perennial Modern Classics, 2006), 155–156.

3. Dave Foreman and Bill Haywood, eds., *Ecodefense: A Field Guide to Monkey-wrenching*, 2nd ed. (Tucson: Ned Ludd Books, 1987), 21.

4. Ibid., 184–185.

5. Amy Irvine, *Trespass: Living at the Edge of the Promised Land* (New York: North Point Press, 2008), 127.

6. "The belief that dead snakes suspended belly-side up on a fence would bring rain is said to have originated before the invention of barbed wire fences in 1874. When laid on a rail or stone fence, *rigor mortis* would invariably cause the snake to twist over onto its belly and thus prevent the 'charm' from being followed to the letter. Barbed wire fences enabled Kansan farmers to penetrate a dorsal segment and so fix the snake in the prescribed position. Thousands of snakes were properly strung on barbed-wire fences, their white-scaled bellies glistening in the brassy sun-glare. But, strange to relate, Nature seldom reacted favorably to the sacrifice of serpents." *Kansas, a Guide to the Sunflower State*, comp. and written by Kansas Federal Writer's Project of the Works Progress Administration for the State of Kansas, American Guide Series (New York: Viking Press, 1939), 101–102.

7. Irvine, *Trespass*, 127–131.

8. Ibid., 336.

9. Ibid., 314–315.

10. Zach Cartwright, "Ranchers at Center of Oregon Standoff Threatened to Wrap Official's Son in Barbed Wire and Drown Him," *U.S. Uncut*, 2016, http://usuncut.com/news/ranchers-oregon-standoff-threatened-official-and-family/ [site discontinued].

11. Edward Abbey, *The Brave Cowboy: An Old Tale in a New Time* (New York: Avon Books, 1992), 11.

12. Ibid., 20.

13. Ibid., 123.

14. Ibid., 219–220.

15. Ellen Meloy, *The Last Cheater's Waltz: Beauty and Violence in the Desert Southwest* (Tucson: University of Arizona Press, 1999), 102.

16. Ibid., 145.

17. Ibid., 14–15.

18. Ibid., 154, 155, 193.

19. James Galvin, *Fencing the Sky* (New York: Picador, 1999), 199, 180, 224.

20. Ibid., 30.

21. Ibid., 47.

22. Ibid., 29, 37.

23. Ibid., 58–59.

24. Ibid., 249.

25. James Galvin, *The Meadow* (New York: Henry Holt, 1992), 5.

26. Mary O'Hara, *My Friend Flicka* (1940; repr., New York: Harper Trophy, 2003), 240, 193.

27. Annie Proulx, *Close Range: Wyoming Stories* (New York: Scribner, 1999), 69, 163, 174, 275.

28. Ibid., 222.

29. Ibid., 231, 234, 219, 225, 236–237, 239.

30. Ibid., 244, 245.

31. Ibid., 19, 20.

32. Ibid., 31, 22, 33, 36, 38.

33. Ibid., 26, 36.

34. Sigmund Freud, *Civilizations and Its Discontents*, trans. James Strachey (New York: W. W. Norton, 2005), 68.

35. Proulx, *Close Range*, 259, 268.

36. Ibid., 275, 276, 279, 284, 261, 269.

Chapter 6

1. Krell, *The Devil's Rope*, 17–19, 152.

2. John Steinbeck, *The Grapes of Wrath* (New York: Viking Press, 1939), 6, 24.

3. Ibid., 39.

4. John Steinbeck, *To a God Unknown* (New York: Penguin Books, 1995), 27, 100, 165.

5. Steinbeck, *Grapes of Wrath*, 47–48.

6. Ibid., 237.

7. *Viva Zapata*, dir. Elia Kazan, screenplay by John Steinbeck (Los Angeles: Twentieth Century Fox, 1952).

8. Steinbeck, *The Grapes of Wrath*, 241, 241–242, 244.

9. Ibid., 389, 393, 518–519.

10. Ibid., 519–520.

11. Ibid., 528.

12. Ibid., 616, 572, 619.

13. Ibid., 31.

14. Ibid., 287, 288–289, 289.

15. Kelly MacPhail argues that the feral nature of the Christians is connected to the descriptions of the monsters of capitalism that prey on the Joads and others. See "'He's—a Kind of a Man': Jim Casy's Spiritual Journey in *The Grapes of Wrath*," in *The Grapes of Wrath: A Re-consideration*, ed. Michael J. Meyer (Amsterdam: Rodopi, 2009), 113–119.

16. Steinbeck, *Grapes of Wrath*, 438, 438, 439, 377.

17. Karl Marx, *Collected Works* (London: Lawrence & Wishart, 1987), 3:175.

18. Georges Bataille, *Visions of Excess: Selected Writings, 1926–1939* (Minneapolis: University of Minnesota Press, 1985), 127.

19. Flannery O'Connor, *Wise Blood*, in *Collected Works: Wise Blood, A Good Man Is Hard to Find, The Violent Bear It Away, Everything That Rises Must Converge, Essays and Letters*, ed. Sally Fitzgerald (New York: Library of America, 1988), 989.

20. In "Unlikely Transfusions: *Wise Blood* and *The Grapes of Wrath*," *Steinbeck Studies* 15, no. 1 (2004): 41–45, John Seelye begins to trace similarities between the two novels by noting that both Tom Joad and Hazel Motes are returning from "regimented and repressive ordeals"—Joad from prison and Motes from the US Army (42). He reminds us that they both return to empty family houses—Joad going on to join a fiercely united family, while Motes has no family left whatsoever. Seelye is primarily interested in the role of religion in both novels, arguing that "Protestant fundamentalism with its earnest, self-invested, self-denying evangelistic zeal . . . holds out hope for society's salvation" (45) in *The Grapes of Wrath* and that the radical Protestantism into which Hazel Motes retreats is destructively solipsistic (44). We see Seelye's reading of evangelical religion in *The Grapes of Wrath* as fundamentally flawed. Steinbeck's novel features a profound renunciation of ecstatic mysticisms, including the version that leads to self-flagellation by barbed wire. And rather than "destructively solipsistic," we see Motes's ascetic practices as historically compatible with O'Connor's own Catholicism.

21. O'Connor, *Collected Works*, 947.

22. Flannery O'Connor, *The Habit of Being: The Letters of Flannery O'Connor*, ed. Sally Fitzgerald (New York: Farrar, Straus & Giroux, 1979), 498.

23. Theodore Ziolkowski, *Fictional Transfigurations of Jesus* (Princeton, NJ: Princeton University Press, 1972), 182–190.

24. O'Connor, *Collected Works*, 9–10, 30, 11. The novel *Wise Blood* was originally published by Harcourt Brace in 1952.

25. Ibid., 126–127. The landlady's surprise is followed by her retreat to avoid potential contact/injury from the physical fence.

26. See, for instance, the spirited interchange between Brian Ingraffia, Susan Srigley, and Ralph Wood, a conversation edited by Susan Srigley: "Penance, Violence, and the Ending of *Wise Blood*: Three Essays in Conversation," *Flannery O'Connor Review* 7 (2009): 78–100.

27. Steinbeck, *The Grapes of Wrath*, 32.

28. Brad Gooch, *Flannery: A Life of Flannery O'Connor* (New York: Little, Brown, 2009), 174, 184, 173, 184.

29. O'Connor, *Collected Works*, 67, 67, 68, 70.

30. Steinbeck, *Grapes of Wrath*, 37.

31. Ibid., 32–33.

32. An image of the music for the song "The Battle Hymn of the Republic" by Julia Ward Howe, first published in the *Atlantic Monthly* 9, no. 52 (February 1862): 145, appears inside the front and back covers of the first edition of *The Grapes of Wrath*.

33. Brian Evenson, "Contagion," in *"Contagion" and Other Stories* (La Grande, OR: Wordcraft of Oregon, 2000), 75.

34. Ibid., 80.

35. Ibid., 82–83.

36. Ibid., 83, 85.

37. Ibid., 87, 88, 90.

38. Ibid., 92.

39. Friedrich Nietzsche, "On Truth and Lie in an Extra-moral Sense," in *The Portable Nietzsche*, ed. Walter Kaufmann (New York: Penguin Books, 1977), 46–47.

40. Evenson, "Contagion," 92–93.

41. Ibid., 93, 96, 96–97.

42. Ibid., 102. The description of the fence cutters is from the famously staged photo of the masked fence cutters on the Brighton Ranch. See Krell, *The Devil's Rope*, 42.

43. Evenson, "Contagion," 102.

44. Nietzsche, cited in Fredric Jameson, *The Prison-House of Language* (Princeton, NJ: Princeton University Press, 1975), i.

45. Evenson, "Contagion," 104.

46. *Glidden Barb-Fence Journal* (1885), 6. We discuss this poem in chapter 3.

47. Steinbeck, *Grapes of Wrath*, 31.

48. Jeff Mann, *A History of Barbed Wire* (Cleveland, OH: Suspect Thoughts Press, 2006), 33.

49. Ibid.

50. Ibid.

51. Ibid.

52. Ibid., 34.

53. Ibid., 35.

54. Ibid., 36. Krell writes about the painful pleasures associated with sex and represented by barbed wire in a sadomasochistic photo by Barbara Nitke, *Gary and wynn IV*. Wynn's barbed wire tattoo represents the actual pain he suffers/enjoys when struck by Gary's whip. See Krell, *The Devil's Rope*, 134.

Chapter 7

1. D'Arcy McNickle, *Wind from an Enemy Sky* (San Francisco: Harper & Row, 1978), 17, 18, 20.

2. Gloria Anzaldúa, *Borderlands/La Frontera: The New Mestiza*, 2nd ed. (San Francisco: Aunt Lute Books, 1987), 81.

3. McCallum and McCallum, *The Wire That Fenced the West*, 203.

4. For a discussion of manipulation of Western photographs to convey diverse meanings, see Martha Sandweiss, "Views and Reviews: Western Art and Western History," in *Under an Open Sky: Rethinking America's Western Past*, ed. William Cronon, George Miles, and Jay Gitlin (New York: W. W. Norton, 1992), 185–202.

5. Curtis similarly doctored photos. A photo whose native authenticity was marred by a clock, for example, was altered to remove the clock. For the original image, see Edward Curtis, *In a Piegan Lodge*, Library of Congress, March 11, ca. 1910, http://www.loc.gov/pictures/item/2002722455. For the altered image, see *In a*

Piegan Lodge, Wikimedia Commons, ca. 1911, https://commons.wikimedia.org/wiki/File:In_a_Piegan_Lodge3.jpg.

6. John P. McGlinn, "Report of Neah Bay Agency," *Fifty-Ninth Annual Report of the Commissioner of Indian Affairs to the Secretary of the Interior* (Washington, DC: Government Printing Office, 1890), 222–225, 223.

7. Ibid., 223.

8. Ibid.

9. Ibid., 224.

10. Evenson, "Contagion," 104.

11. Gerald Robert Vizenor, *Manifest Manners: Narratives on Postindian Survivance* (Lincoln: University of Nebraska Press, 1999), 13.

12. A Company for Breeding Horses on the Crow Indian Reservation, Montana, the Survey and Allotment of Indian Lands within the Limits of the Crow Indian Reservation, Montana. Hearings before the Committee on Indian Affairs, United States Senate on the Bill S. 2087 to Incorporate a Company for Breeding Horses on the Crow Indian Reservation, Montana, and for other Purposes and on the Bill S. 2963 for the Survey and Allotment of Indian Lands Now Embraced within the Limits of the Crow Indian Reservation. Senate Doc. No. 445, 60th Cong., 1908, 8, 15.

13. Ibid., 249.

14. Ibid., 531.

15. Ibid., 588.

16. Ibid., 654.

17. Ibid., 654–672. See the account of Grey's influence on the Crows in Frederick E. Hoxie, *Parading through History: The Making of the Crow Nation in America 1805–1935* (Cambridge: Cambridge University Press, 1996), 238–247.

18. James Welch, *Winter in the Blood* (New York: Penguin Books, 2008), 1–2.

19. Bennett, *Vibrant Matter*.

20. *Skins*, dir. Chris Eyre (Los Angeles: First Look Pictures, 2002).

21. *Barking Water*, dir. Sterlin Harjo (Tulsa, OK: Indion Entertainment Group, 2009).

22. Robert J. Bigart and Clarence Woodcock, eds., *In the Name of the Salish and Kootenai Nation: The 1855 Hell Gate Treaty and the Origin of the Flathead Indian Reservation* (Pablo, MT: Salish Kootenai College Press, 1996), 122–125, 142.

23. *Indian Affairs: Laws and Treaties*, vol. 1, *Laws*, comp. and ed. Charles J. Kappler (Washington, DC: Government Printing Office, 1904), chap. 308, June 5, 1872, 17 Stat., https://babel.hathitrust.org/cgi/pt?id=osu.32435021181110;view=1up;seq=9. The 1889 Allotment Act (chap. 39, March 2, 25 Stat., 871) permitted Indian allotments in Bitter Root Valley to be sold. And the 1904 Flathead Allotment Act ([H.R. 12231; Public Law No. 159] chap. 1495, April 23, 1904) was called "an Act for the survey and allotment of lands now embraced within the limits of the Flathead Indian Reservation, in the state of Montana, the sale and disposal of all surplus lands after allotment." The 1971 US Indian Court of Claims ruled this act a breach of the Hellgate Treaty.

24. On September 4, 2015, the Confederated Salish and Kootenai Tribes bought the Kerr Hydroelectric project from Northwestern Energy for $18.3 million and transferred ownership to the tribally owned Energy Keepers. This was the first Indian nation/tribe to buy a dam on its lands. Additionally, the tribe has bought back about 70 percent of original reservation lands. See Roy Bigcrane and Thompson Smith's documentary about pre- and post-dam history of the Flathead Reservation, *The Place of the Falling Waters*, Sacred Land Film Project (Pablo, MT: Salish Kootenai College Media Center, 1990).

25. Dorothy R. Parker, *Singing an Indian Song: A Biography of D'Arcy McNickle* (Lincoln: University of Nebraska Press, 1992), 113.

26. McNickle, *Wind from an Enemy Sky*, 222.

27. Ibid., 229.

28. Ibid., 121.

29. Ibid., 234. McNickle participated in negotiations for the return of a sacred bundle in 1937–1938, which belonged to the Water Buster Clan of the Gros Ventre Tribe of North Dakota; the Feather Boy bundle was based loosely on this. See Louis Owens, "Afterword," in the reprint of *Wind from an Enemy Sky* (Albuquerque: University of New Mexico Press, 1988), 260.

30. McNickle, *Wind from an Enemy Sky,* 2, 125.

31. Shari M. Huhndorf and Scott L. Pratt, "Cultural Cartographies: The Logic of Domination and Native Cultural Survival," *Journal of Speculative Philosophy* 14, no. 4 (2000): 269.

32. McNickle, *Wind from and Enemy Sky*, 256.

33. Parker, *Singing an Indian Song*, 232.

34. D'Arcy McNickle, *Native American Tribalism: Indian Survivals and Renewals* (New York: Oxford University Press, 1973), 10.

35. Huhndorf and Pratt, "Cultural Cartographies," 283. For a fuller discussion of competing mind maps in the novel and of the two young men as hybrid characters representing a new kind of map, see ibid., 282–283.

36. Paula Gunn Allen, *Off the Reservation: Reflections on Boundary-Busting, Border Crossing Loose Canons* (Boston: Beacon Press, 1998), 2, 6, 6, 7.

37. Ibid., 7.

38. Homi Bhabha, *The Location of Culture* (London: Routledge Classics, 2004), 2–3.

39. Richard Frost, *The Railroad and the Pueblo Indians* (Salt Lake City: University of Utah Press, 2016), 139–140.

40. Ibid., 142, 142, 157.

41. Robin Cohen, "Of Apricots, Orchids, and Wovoka: An Interview with Leslie Marmon Silko," in *Leslie Marmon Silko's* Ceremony, ed. Allan Chavkin (New York: Oxford University Press, 2002), 258.

42. Leslie Marmon Silko, *Ceremony* (New York: Penguin Books, 1977), 2.

43. Ibid., 188, 27, 79.

44. Ibid., 44.

45. *Glidden Barb-Fence Journal* (1885), 15.

46. Silko, *Ceremony*, 126.

47. Ibid., 186.

48. Ibid., 187–188.

49. Ibid., 189.

50. Ibid., 190–191.

51. Vizenor, *Manifest Manners*, 12.

52. Silko, *Ceremony*, 201–202.

53. Paula Gunn Allen, "Iyani: It Goes This Way," in *The Remembered Earth: An Anthology of Contemporary Native American Literature*, ed. Geary Hobson (Albuquerque: University of New Mexico Press, 1981), 191.

54. Silko, *Ceremony*, 246, 246, 191, 247.

55. Ibid., 251, 253.

56. Ibid., 258.

57. Bhabha, *The Location of Culture*, 1.

58. Ibid., 2.

59. Anzaldúa, *Borderlands/La Frontera*, 23–25.

60. For a fine history of the US-Mexico border, see Rachel St. John, *Line in the Sand* (Princeton, NJ: Princeton University Press, 2011).

61. Anzaldúa, *Borderlands/La Frontera*, 24–25.

62. "She puts history through a sieve, winnows out the lies, looks at the forces that we as a race, as women, have been a part of. . . . This step is a conscious rupture with all oppressive traditions of all cultures and religions. She communicates that rupture, documents the struggle. She reinterprets history and, using new symbols, she shapes new myths." Ibid., 104.

63. Ibid., 25. In the introduction we explored a similar juxtaposition of wind and barbed wire in Amy Clampitt's poem "The Woodlot."

64. For a rich overview of the use of barbed wire in artworks about the US-Mexico border region, see María Herrera-Sobek, "Barbed Wire Iconography and Aesthetic Activism: The Borderlands, Mexican Immigration and Chicana/o Art," in *International Perspectives on Chicana/o Studies: "This World Is My Place,"* ed. Catherine Leen and Niamh Thornton (New York: Routledge, 2013), 150–167.

65. John D'Agata, *Halls of Fame* (Minneapolis: Graywolf Press, 2001), 109–114.

Conclusion

1. Brown, *A Sense of Things*, 4.

2. David Lee, *The Porcine Canticles* (Port Townsend, WA: Copper Canyon Press, 1984), 16.

3. Ibid., 15, 14.

4. Nauman, *Setting a Good Corner*.

5. Terry Tempest Williams, *Refuge* (New York: Vintage, 1992), 288.

6. William Faulkner, *A Fable* (New York: Vintage International, 2011), 142, 424.

7. Ryszard Kapuscinski, *Imperium*, trans. Klara Glowczewska (New York: Vintage International, 1995), 20.

8. Ibid., 21–22.

9. Ibid., 85, 202, 86.

10. Executive Order 9066 authorized the designation of military areas in the United States "from which any or all persons may be excluded" for "protection against espionage and against sabotage." For complete text of order, see "Document for February 19th: Executive Order 9066: Resulting in the Relocation of Japanese," National Archives, February 19, 1942, http://www.archives.gov/historical-docs/todays-doc/?dod-date=219.

11. John Arcudi, *Barb Wire, Steel Harbor's Babe on Wheels, Devil in the Dark*, no. 1 (Los Angeles: Dark Horse Comics, 1994), cover. Adam Warren was the penciler and did the ink for the cover.

12. Pamela Anderson plays Barb Wire in the film of the same name. *Barb Wire*, dir. David Hogan (Los Angeles: Propaganda Films and Dark Horse Entertainment, 1996).

13. Bo Diddley (a.k.a. Ellas McDaniel), "Who Do You Love?" (Chicago: Checkers Records, 1956).

14. Description from a review of the exhibit *Dalí & Film* at the Tate Modern by Serena Davis, "Harpo and Dalí: A Surreal Double Act," *The Telegraph*, May 26, 2007, http://www.telegraph.co.uk./culture/art/3665391/Harpo-and-Dali-a-surreal-double-act.html.

15. Yehuda Amichai, *Open Closed Open*, trans. Chana Bloch and Chana Kronfeld (Tel Aviv: Schocken, 1998), 67. We quote a similar but unpublished translation by Edwin Firmage.

16. "Ein Yahav," Wikipedia, January 22, 2017, https://en.wikipedia.org/wiki/Ein_Yahav.

17. Wislawa Szymborska, "Starvation Camp near Jaslo," in *Poems, New and Collected* (New York: Mariner Books, 2000), 42.

BIBLIOGRAPHY

Archives/Special Collections

American Steel & Wire Company Papers. Baker Business Historical Collections. Harvard University. Boston, Massachusetts.

Barbed Wire Remedies Bottle Collection. Kansas Barbed Wire Museum. La Crosse, Kansas.

California Room. California State Library. Sacramento, California.

Ellwood Family Collection, Earl Hayter Collection, Henry De Rossett McCallum and Frances Tarlton McCallum Papers, and Henry W. Putnam Papers. University of Wyoming. American Heritage Center. Laramie, Wyoming.

Ellwood House Museum. DeKalb, Illinois.

L. Tom Perry Special Collections. Harold B. Lee Library. Brigham Young University. Provo, Utah.

Regional History Center. Northern Illinois University. DeKalb, Illinois.

Western History/Genealogy Department. Denver Public Library. Denver, Colorado.

Government Documents

A Company for Breeding Horses on the Crow Indian Reservation, Montana, the Survey and Allotment of Indian Lands within the Limits of the Crow Indian Reservation, Montana. Hearings before the Committee on Indian Affairs, United States Senate on the Bill S. 2087 to Incorporate a Company for Breeding Horses on the Crow Indian Reservation, Montana, and for Other Purposes and on the Bill S. 2963 for the Survey and Allotment of Indian Lands Now Embraced within the Limits of the Crow Indian Reservation. Senate Doc. No. 445. 60th Cong. 1908.

"Document for February 19th: Executive Order 9066: Resulting in the Relocation of Japanese." National Archives, February 19, 1942. http://www.archives.gov/historical-docs/todays-doc/?dod-date=219.

Gammel, H. P. N., comp. *The Laws of Texas 1822–1897.* Austin: Gammel, 1898.

Indian Affairs: Laws and Treaties. Vol. 1, *Laws.* Compiled and edited by Charles J. Kappler. Washington: Government Printing Office, 1904. Chap. 308, June 5, 1872, 17 Stat. https://babel.hathitrust.org/cgi/pt?id=osu.32435021181110;view=1up;seq=9.

Journal of the House of Representatives of the State of Vermont. Biennial Session, 1880. Montpelier, VT: Freeman Steam Printing House and Bindery, 1881.

Journal of the Senate of the State of Vermont. Biennial Session, 1880. Montpelier, VT: Freeman Steam Printing House and Bindery, 1881.

Justia US Supreme Court. "The Barbed Wire Patent." 143 US 275 (1892), 143 US 292. https://supreme.justia.com/cases/federal/us/143/275/case.html.

Letter from the Secretary of the Interior, Transmitting, in Answer to Senate Resolution of February 14, Letter of the Commissioner of the General Land Office, on the Subject of Unauthorized Fencing of Public Lands. Exec. Doc. No. 127. 48th Cong., 1st sess. March 14, 1884.

McGlinn, John P. "Report of Neah Bay Agency." In *Fifty-Ninth Annual Report of the Commissioner of Indian Affairs to the Secretary of the Interior.* Washington, DC: Government Printing Office, 1890.

"Proclamation No. 271—Ordering the Immediate Removal of Any and Every Unlawful Enclosure of Public Land." August 7, 1885. The American Presidency Project. http://www.presidency.ucsb.edu/ws/?pid=71889.

Public Lands. Unlawful Inclosures or Occupancy; Obstructing Settlement or Transit, U.S.C. 1061, Title 43 (1885).

Sherman Antitrust Act. Sec. 2, 51st Cong., 1st sess., 1890. http://www.stern.nyu.edu/networks/ShermanClaytonFTC_Acts.pdf.

Transactions of the Iowa State Horticultural Society for 1888. Twenty-Third Annual Session held at Des Moines, January 15, 16, 17, 1889. Also Proceedings of the Western, Eastern and Northern Societies, with Summary of Other Local Societies for the Year 1888. Vol. 23. Edited by Geo. Van Houten. Des Moines, IA: Geo. E. Roberts, State Printer, 1889.

United States Patent and Trademark Office. 1867–1902. Alexandria, VA.

Periodicals

Abbeville Press and Banner (Abbeville, SC)
American Agriculturalist (New York, NY)
Arizona Sentinel (Yuma, Arizona Territory)
Austin Weekly Statesman (Austin, TX)
Barbour County Index (Medicine Lodge, KS)
Belmont Chronicle (St. Clairsville, OH)
Bismarck Tribune (Bismarck, Dakota Territory)
Breckenridge News (Cloverport, KY)
Breeder's Gazette (Chicago, IL)
Brenham Weekly Banner (Brenham, TX)
Burlington Weekly Free Press (Burlington, VT)
Butler Citizen (Butler, PA)
Butler Weekly Times (Butler, MO)
Caledonian (St. Johnsbury, VT)
Canton Advocate (Canton, Dakota Territory)
Century Magazine, Illustrated Monthly (New York, NY)
Chariton Courier (Keytesville, MO)
Chicago Daily Tribune (Chicago, IL)

Connecticut Farmer (Hartford, CT)
Cultivator & Country Gentleman (Albany, NY)
Daily Globe (St. Paul, MN)
Daily News (Denver, CO)
Dallas Daily Herald (Dallas, TX)
Democratic Northwest (Napoleon, OH)
Dodge City Times (Dodge City, KS)
Edgefield Advertiser (Edgefield, SC)
Elk County Advocate (Ridgway, PA)
Emporia News (Emporia, KS)
Emporia Weekly News (Emporia, KS)
Evening Star (Washington, DC)
Fayetteville Observer (Fayetteville, TN)
Fort Worth Daily Gazette (Fort Worth, TX)
Galveston Daily News (Galveston, TX)
Green-Mountain Freeman (Montpelier, VT)
Hartford Daily Courant (Hartford, CT)
Helena Weekly Herald (Helena, Montana Territory)
Holt County Sentinel (Oregon, MO)
Idaho Semi-Weekly World (Idaho City, Idaho Territory)
Iola Register (Iola, KS)
Iron Age (New York, NY)
Jackson Standard (Jackson, OH)
Jamestown Alert (Jamestown, Dakota Territory)
Lancaster Daily Intelligencer (Lancaster, PA)
Las Vegas Daily Gazette (Las Vegas, New Mexico Territory)
Leavenworth Weekly Times (Leavenworth, KS)
Lewiston Teller (Lewiston, North Idaho Territory)
Lexington Weekly Intelligencer (Lexington, MO)
Los Angeles Daily Herald (Los Angeles, CA)
Marshall County Republican (Plymouth, IN)
McCook Weekly Tribune (McCook, NE)
The Meridional (Abbeville, LA)
Mexico Weekly Ledger (Mexico, MO)
Middlebury Register (Middlebury, VT)
Milan Exchange (Milan, TN)
Mower County Transcript (Austin, MN)
National Live-Stock Journal-Weekly Edition (Chicago, IL)
National Tribune (Washington, DC)
Nebraska Advertiser (Brownville, NE)
New Bloomfield, PA Times (New Bloomfield, PA)
New North-West (Deer Lodge, Montana Territory)
New Orleans Daily Democrat (New Orleans, LA)

New Ulm Weekly Review (New Ulm, MN)
New York Times (New York, NY)
News and Citizen (Morrisville and Hyde Park, VT)
News and Herald (Winnsboro, SC)
Northern Pacific Farmer (Wadena, MN)
Omaha Daily Bee (Omaha, NE)
Ottawa Free Trader (Ottawa, IL)
Our Dumb Animals (Boston, MA)
Pascagoula Democrat-Star (Pascagoula, MS)
Phillipsburg Herald (Phillipsburg, KS)
Prairie Farmer (Chicago, IL)
Press and Daily Dakotaian (Yankton, Dakota Territory)
Red Cloud Chief (Red Cloud, NE)
Richmond Democrat (Richmond, MO)
River Press (Fort Benton, Montana Territory)
Rock Island Argus (Rock Island, IL)
Rocky Mountain Husbandman (White Sulphur Springs, Montana Territory)
Sacramento Daily Record-Union (Sacramento, CA)
Saline County Journal (Salina, KS)
San Saba News (San Saba, TX)
Scribner's Magazine (New York, NY)
Sedalia Weekly Bazoo (Sedalia, MO)
Semi-Weekly Interior Journal (Stanford, KY)
Shenandoah Herald (Woodstock, VA)
Southern Cultivator and Dixie Farmer (Augusta, GA)
St. Landry Democrat (Opelousas, LA)
The Times (Owosso, MI)
Times and Democrat (Orangeburg, SC)
True Northerner (Paw Paw, MI)
Vermont Phoenix and Record and Farmer (Brattleboro, VT)
Vermont Watchman & State Journal (Montpelier, VT)
Waco Daily Examiner (Waco, TX)
Weekly Arizona Miner (Prescott, Arizona Territory)
Weekly Chillicothe Crisis (Chillicothe, MO)
Weekly Clarion (Jackson, MI)
Weekly Democratic Statesman (Austin, TX)
Weekly Kansas Chief (Troy, KS)
Weekly Miner (Butte, Montana Territory)
Weekly Register (Point Pleasant, WV)
Wheeling Daily Intelligencer (Wheeling, WV)
Wichita City Eagle (Wichita, KS)
Worthington Advance (Worthington, MN)

Trade Manuals and Almanacs

Barb Fence: Its Utility, Efficiency and Economy: A Book for the Farmer, the Gardener and the Country Gentleman. Worcester, MA: Lucius P. Goddard, 1876.

Barb Fence: Its Utility, Efficiency and Economy: A Book for the Farmer, the Gardener and the Country Gentleman. Worcester, MA: Noyes, Snow, 1878.

The Barb Fence Regulator. DeKalb, IL: Jacob Haish, 1877(?)–1890(?).

Barb Fencing. Statement of the Washburn & Moen Manufacturing Co., before the Committee on Agriculture of the General Assembly of Connecticut, Thursday, February 9, 1880. Worcester, MA: Press of Snow, Woodman, 1880.

Barb Wire Fencing: A Statement of the Washburn & Moen Manufacturing Co. to the Legislature of New Hampshire, July 12, 1881. Manchester, NH: John B. Clarke, 1881.

"Barbed Fence Wire." J. W. Gates Circular. St. Louis, MO(?): N.p., 1882.

The Celebrated Baker Perfect Barb Wire, Pocket Record. Chicago, IL: Consolidated Steel & Wire, 1895.

The DeKalb County Manufacturer, with Illustrations and Historical Sketches of DeKalb County Inventions. DeKalb, IL: Chronicle Press and Bindery, 1882.

Do You Use Barbed Wire? Try the Only Perfect Wire Made: Freeman's Diamond. East St. Louis, MO: Freeman Wire & Iron, n.d.

"The Eccentricities of Jacob Haish." Flier. N.p: Thorn Wire Hedge Company, n.d.

The Fence: A Compilation of Facts, Figures and Opinions from National and State Agricultural Reports, Agricultural Journals, and the Public Press, for the Past Sixty Years, (Presented in Their Original Form,) Showing the Waste and Need of Fencing Material, the Cost of Fences, Their Waste and Wear in the United States, Relation to Other Farm Property, etc., etc. and a Compilation of Statutory and Judicial Facts and Decisions in the State and Federal Courts, Relating to Railway Fences and Fence Laws, with the Early History of Enclosures. Worcester, MA: Press of Noyes, Snow, 1879.

The Fence Problem in the United States as Related to General Husbandry and Sheep Raising. Facts and Statistics from Authoritative Sources, with a View of Fence Laws and Customs. Worcester, MA: Washburn & Moen Manufacturing, 1882.

The Glidden Barb-Fence Journal. DeKalb, IL: I. L. Ellwood 1881–1888.

The Glidden Pocket Compendium for 1882. DeKalb, IL: I. L. Ellwood, 1882.

Glidden Patent Steel Barb Wire Fencing. Worcester, MA: Washburn & Moen, n.d.

Glidden Patent Two Point and Four Point Barb Wire Fencing: Cheap, Staunch, Indispensable. Worcester, MA: Washburn & Moen, n.d.

Haish, Jacob. *A Reminiscent Chapter from the Unwritten History of Barb Wire prior to and Immediately following the Celebrated Decision of Judge Blodgett, December 15, 1880*. N.p., 188(?).

Haller Catalogue, Health Guide, & Almanac, 1908. Blair, NE: Haller Proprietary, 1908.

A Manual of the Fence: Comprising the Fence Complaint of the Past, and the Fence Remedy Assured in Steel Barb Fencing. Worcester, MA: Washburn & Moen, 1881.

Pacific Steel & Wire Company: Fencing for All Purposes. San Francisco, ca. 1897.

Sears, Roebuck & Co., Cheapest Supply House on Earth. Catalogue No. 109. Chicago, IL: Sears, Roebuck, Fall 1899.

Steel Barb Fencing, the World's Fence, Everywhere Adapted to All the Uses of Husbandry in All Climates and under All Exposures, "the Perfect Fence." Worcester, MA: Press of Snow, Woodman, 1880.

Books and Articles

Abbey, Edward. *The Brave Cowboy: An Old Tale in a New Time*. New York: Avon Books, 1992.

———. *The Monkey Wrench Gang*. New York: Harper Perennial Modern Classics, 2006.

Allen, Paula Gunn. "Iyani: It Goes This Way." In The *Remembered Earth: An Anthology of Contemporary Native American Literature*, edited by Geary Hobson, 191–193. Albuquerque: University of New Mexico Press, 1981.

———. *Off the Reservation: Reflections on Boundary-Busting, Border Crossing Loose Canons*. Boston: Beacon Press, 1998.

Amichai, Yehuda. *Open Closed Open*. Translated by Chana Bloch and Chana Kronfeld. Tel Aviv: Schocken, 1998.

Anzaldúa, Gloria. *Borderlands/La Frontera: The New Mestiza*. 2nd ed. San Francisco: Aunt Lute Books, 1987.

Appadurai, Arjun, ed. *The Social Life of Things: Commodities in Cultural Perspective*. Cambridge: Cambridge University Press, 1986.

Arcudi, John. *Barb Wire, Steel Harbor's Babe on Wheels, Devil in the Dark*, no. 1. Los Angeles: Dark Horse Comics, 1994.

Bataille, Georges. *Visions of Excess: Selected Writings, 1926–1939*. Minneapolis: University of Minnesota Press, 1985.

Bell, Bob Boze. "Classic Gunfights: To the Last Man." *True West*, 2017. http://www.truewestmagazine.com/classic-gunfights-last-man/.

Bennett, Jane. *Vibrant Matter: A Political Ecology of Things*. Durham, NC: Duke University Press, 2010.

Bhabha, Homi. *The Location of Culture*. London: Routledge Classics, 2004.

Bigart, Robert J., and Clarence Woodcock, eds. *In the Name of the Salish and Kootenai Nation: The 1855 Hell Gate Treaty and the Origin of the Flathead Indian Reservation*. Pablo, MT: Salish Kootenai College Press, 1996.

Boskin, Joseph. *Sambo: Rise & Demise of an American Jester*. Oxford: Oxford University Press, 1986.

Bourcier, Paul G. "'In Excellent Order': The Gentleman Farmer Views His Fences, 1790–1860." *Agricultural History* 58, no. 4 (1984): 546–564.

Brown, Bill. *A Sense of Things: The Object Matter of American Literature*. Chicago: University of Chicago Press, 2003.

Byrd, Kathryn E. "Barbed Wire: Its Origin and Effects." Master's thesis, Southwest Texas State Teachers College, 1944.

Cartwright, Zach. "Ranchers at Center of Oregon Standoff Threatened to Wrap Official's Son in Barbed Wire and Drown Him." *U.S. Uncut*, 2016. http:

//usuncut.com/news/ranchers-oregon-standoff-threatened-official-and-family/. Website subsequently shut down.

Clampitt, Amy. *The Collected Poems of Amy Clampitt*. New York: Knopf, 1999.

———. *The Kingfisher*. New York: Knopf, 1982.

Cohen, Robin. "Of Apricots, Orchids, and Wovoka: An Interview with Leslie Marmon Silko." In *Leslie Marmon Silko's* Ceremony, edited by Allan Chavkin, 257–264. New York: Oxford University Press, 2002.

Comer, Krista. *Landscapes of the New West: Gender and Geography in Contemporary Women's Writing*. Chapel Hill: University of North Carolina Press, 1999.

Costello, Matthew. "'I Didn't Expect to Find Any Fences around Here': Cultural Ambiguity and Containment in *Shane*." *Journal of American Culture* 27, no. 3 (2004): 261–270.

Cushman, Gregory T. *Guano and the Opening of the Pacific World: A Global Ecological History*. New York: Cambridge University Press, 2013.

D'Agata, John. *Halls of Fame*. Minneapolis: Graywolf Press, 2001.

Davis, Mollie E. Moore. *The Wire Cutters*. College Station: Texas A&M University Press, 1997.

Davis, Serena. "Harpo and Dalí: A Surreal Double Act." *The Telegraph*, May 26, 2007. http://www.telegraph.co.uk./culture/art/3665391/Harpo-and-Dali-a-surreal-double-act.html.

De Bevere, Maurice [Morris] & René Goscinny. *A Lucky Luke Adventure: Barbed Wire on the Prairie*. Translated by Luke Spear. Ashford, Kent, UK: Cinebook, 2007.

Deetz, James. *In Small Things Forgotten: The Archaeology of Early North America Life*. Exp. and rev. New York: Anchor Books, 1996.

Diddley, Bo. "Who Do You Love?" Chicago: Checkers Records, 1956.

Donoghue, Denis. *Metaphor*. Cambridge, MA: Harvard University Press, 2014.

Ehrlich, Gretel. "The Solace of Open Spaces." In *New West Reader: Essays on an Ever-Evolving Frontier*, edited by Philip Connors, 90–104. New York: Nation Books, 2005.

Evans, Sterling. *Bound in Twine: The History and Ecology of the Henequen-Wheat Complex for Mexico and the American and Canadian Plains, 1880–1950*. College Station: Texas A&M University Press, 2007.

Evenson, Brian. *"Contagion" and Other Stories.* La Grande, OR: Wordcraft of Oregon, 2000.

Faulkner, William. *A Fable*. New York: Vintage International, 2011.

Foreman, Dave, and Bill Haywood, eds. *Ecodefense: A Field Guide to Monkeywrenching*. 2nd ed. Tucson: Ned Ludd Books, 1987.

Freud, Sigmund. *Civilizations and Its Discontents*. Translated by James Strachey. New York: W. W. Norton, 2005.

Frost, Richard. *The Railroad and the Pueblo Indians*. Salt Lake City: University of Utah Press, 2016.

Galvin, James. *Fencing the Sky*. New York: Picador, 1999.

———. *The Meadow*. New York: Henry Holt, 1992.

Gard, Wayne. "The Fence-Cutters." *Southwestern Historical Quarterly* 51, no. 1 (1947): 1–15.

Geisst, Charles R. *Monopolies in America: Empire Builders and Their Enemies from Jay Gould to Bill Gates*. New York: Oxford University Press, 2000.

Glover, Jack. *"Bobbed" Wire: An Illustrated Guide to the Identification and Classification of Barbed Wire*. Wichita Falls, TX: Terry Bros., 1966.

Gooch, Brad. *Flannery: A Life of Flannery O'Connor.* New York: Little, Brown, 2009.

Goodwyn, Lawrence. *Democratic Promise: The Populist Moment in America*. Oxford: Oxford University Press, 1976.

Grey, Zane. *The Drift Fence*. New York: Grosset & Dunlap, 1933.

Hagemeier, Harold. *The Barbed Wire Identification Encyclopedia*. 5th ed. Kearney, NE: Morris Press, 2010.

———. *Supplement to Edition Five of Barbed Wire Identification Encyclopedia*. Kearney, NE: Morris Press, 2012.

Handley, William R. "Wister's Omniscience and Omissions." In *Reading* The Virginian *in the New West*, edited by Melody Graulich and Stephen Tatum, 39–71. Lincoln: University of Nebraska Press, 2003.

Hayter, Earl W. "Barbed Wire Fencing—a Prairie Invention: Its Rise and Influence in the Western States." *Agricultural History* 13, no. 4 (1939): 189–207.

———. "The Fencing of Western Railways." *Agricultural History* 19, no. 3 (1945): 163–167.

Herner, Charles. *The Arizona Rough Riders*. Tucson: University of Arizona Press, 1970.

Herrera-Sobek, María. "Barbed Wire Iconography and Aesthetic Activism: The Borderlands, Mexican Immigration, and Chicana/o Art." In *International Perspectives on Chicana/o Studies: "This World Is My Place,"* edited by Catherine Leen and Niamh Thornton, 150–167. New York: Routledge, 2013.

Howe, Julia Ward. "The Battle Hymn of the Republic." *Atlantic Monthly* 9, no. 52 (February 1862): 145.

Hoxie, Frederick E. *Parading through History: The Making of the Crow Nation in America 1805–1935*. Cambridge: Cambridge University Press, 1996.

Huhndorf, Shari M., and Scott L. Pratt. "Cultural Cartographies: The Logic of Domination and Native Cultural Survival." *Journal of Speculative Philosophy* 14, no. 4 (2000): 268–285.

Ingraffia, Brian, Susan Srigley, and Ralph Wood. "Penance, Violence, and the Ending of *Wise Blood*: Three Essays in Conversation." *Flannery O'Connor Review* 7 (2009): 78–100.

Irvine, Amy. *Trespass: Living at the Edge of the Promised Land.* New York: North Point Press, 2008.

Jameson, Fredric. *The Prison-House of Language*. Princeton, NJ: Princeton University Press, 1975.

Jay, Robert. *The Trade Card in Nineteenth-Century America*. Columbia: University of Missouri Press, 1987.

Jones, Robert Huhn. "Three Days of Violence: The Regulators of the Rock River Valley." *Journal of the Illinois State Historical Society* 59, no. 2 (1966): 131–142.

Jordan, Terry G. *German Seed in Texas Soil: Immigrant Farmers in Nineteenth-Century Texas*. Austin: University of Texas Press, 1966.

Joyce, James. *Ulysses*. Edited by Hans Walter Gabler. New York: Vintage Books, 1986.

Judd, Cameron. *Devil Wire/Brazos*. New York: St. Martin's Paperbacks, 2008.

Kant, Immanuel. *Critique of Judgement*. Translated by James Creed Meredith. Oxford: Oxford University Press, 2007.

Kansas, a Guide to the Sunflower State. Compiled and by Kansas Writer's Federal Writers Project of the Work Projects Administration for the State of Kansas. American Guide Series. New York: Viking Press, 1939.

Kaplan, Justin. *Mr. Clemens and Mark Twain: A Biography*. New York: Simon & Schuster, 2006.

Kapuscinski, Ryszard. *Imperium*. Translated by Klara Glowczewska. New York: Vintage International, 1995.

Kelton, Elmer. *Barbed Wire*. New York: Forge Books, 2006.

———. *Shotgun and Six Bits a Day*. New York: Forge Books, 2014.

Krell, Alan. *The Devil's Rope: A Cultural History of Barbed Wire*. London: Reaktion Books, 2002.

Laird, Pamela Walker. *Advertising Progress: American Business and the Rise of Consumer Marketing*. Baltimore: Johns Hopkins University Press, 1998.

L'Amour, Louis. *Kiowa Trail*. New York: Bantam Books, 2009.

Lears, Jackson. *Fables of Abundance: A Cultural History of Advertising in America*. New York: Basic Books, 1994.

Lee, David. *The Porcine Canticles*. Port Townsend, WA: Copper Canyon Press, 1984.

Limerick, Patricia Nelson. *The Legacy of Conquest: The Unbroken Past of the American West*. New York: W. W. Norton, 1987.

Limerick, Patricia Nelson, Clyde A. Milner II, and Charles E. Rankin, eds. *Trails: Toward A New Western History*. Lawrence: University Press of Kansas, 1991.

Lincoln, Elliot. *Rhymes of a Homesteader*. Boston: Houghton Mifflin, 1920.

Liu, Joanne S. *Barbed Wire: The Fence That Changed the West*. Missoula, MT: Mountain Press, 2009.

London, Jack. *The Road*. New York: Macmillan, 1907.

Luebke, Frederick. *Ethnicity on the Great Plains*. Lincoln: University of Nebraska Press, 1980.

MacPhail, Kelly. "'He's—a Kind of a Man': Jim Casy's Spiritual Journey in *The Grapes of Wrath*." In *The Grapes of Wrath: A Re-consideration*, edited by Michael J. Meyer, 123–153. Amsterdam: Rodopi Press, 2009.

Mann, Jeff. *A History of Barbed Wire*. Cleveland, OH: Suspect Thoughts Press, 2006.

Marcellus, Jane. "Nervous Women and Noble Savages: The Romanticized 'Other' in Nineteenth-Century US Patent Medicine Advertising." *Journal of Popular Culture* 41, no. 5 (2008): 784–808.

Marchand, Roland. *Advertising the American Dream: Making Way for Modernity, 1920–1940*. Berkeley: University of California Press, 1985.

Marx, Karl. *Collected Works*. Vol. 3. London: Lawrence & Wishart, 1987.

Marx, Leo. *The Machine in the Garden*. Oxford: Oxford University Press, 2000.

McCallum, Henry D., and Frances T. McCallum. *The Wire That Fenced the West*. Norman: University of Oklahoma Press, 1965.

McClure, C. Boone. "History of the Manufacture of Barbed Wire." *Panhandle Plains Historical Review* 23 (1958): 1–114.

McFadden, Joseph M. "Monopoly in Barbed Wire: The Formation of the American Steel and Wire Company." *Business History Review* 52, no. 4 (1978): 465–489.

McMurry, Sally Ann. *Families and Farmhouses in Nineteenth-Century America: Vernacular Design and Social Change*. Oxford: Oxford University Press, 1988.

McNickle, D'Arcy. *Native American Tribalism: Indian Survivals and Renewals*. New York: Oxford University Press, 1973.

———. *Wind from an Enemy Sky*. San Francisco: Harper & Row, 1978.

Meloy, Ellen. *The Last Cheater's Waltz: Beauty and Violence in the Desert Southwest*. Tucson: University of Arizona Press, 1999.

Miller, Arthur G., ed. *In the Eye of the Beholder: Contemporary Issues in Stereotyping*. New York: Praeger, 1982.

Nauman, Bruce. "Setting a Good Corner: Bruce Nauman." ART21, November 2011. http://www.ART21.org/read/bruce-nauman-setting-a-good-corner.

Nelson, Cary, and Michael Chasar. "American Advertising: A Poem for Every Product." In *The Oxford History of Popular Print Culture*, edited by Christine Bold, 6:133–167. Oxford: Oxford University Press, 2012.

Netz, Reviel. *Barbed Wire: An Ecology of Modernity*. Middletown, CT: Wesleyan University Press, 2004.

Nietzsche, Friedrich. "On Truth and Lie in an Extra-moral Sense." In *The Portable Nietzsche*, edited by Walter Kaufmann, 42–47. New York: Penguin Books, 1977.

Obata, Chiura. *Chiura Obata's Topaz Moon: Art of the Internment Camps*. Edited by Kimi Kodani Hill. Berkeley, CA: Heyday Books, 2000.

O'Connor, Flannery. *The Habit of Being: The Letters of Flannery O'Connor*. Edited by Sally Fitzgerald. New York: Farrar, Straus & Giroux, 1979.

———. *Wise Blood*. In *Collected Works: Wise Blood, A Good Man Is Hard to Find, The Violent Bear It Away, Everything That Rises Must Converge, Essays and Letters*, edited by Sally Fitzgerald, 3–131. New York: Library of America, 1988.

O'Hara, Mary. *My Friend Flicka*. New York: Harper Trophy, 2003.

Ohmann, Richard. *Selling Culture: Magazines, Markets, and Class at the Turn of the Century*. New York: Verso, 1996.

Oles, Thomas. *Walls: Enclosure and Ethics in the Modern Landscape*. Chicago: University of Chicago Press, 2015.

Oliver, Mary. *New and Selected Poems*. Boston: Beacon Press, 1992.

Owens, Louis. "Afterword." In *Wind from an Enemy Sky*, by D'Arcy McNickle, 260. 1978. Repr., Albuquerque: University of New Mexico Press, 1988.

Panofsky, Erwin. *Studies in Iconology*. New York: Oxford University Press, 1939.

Parker, Dorothy R. *Singing an Indian Song: A Biography of D'Arcy McNickle*. Lincoln: University of Nebraska Press, 1992.

Peterson, Fred W. "Vernacular Building and Victorian Architecture: Midwestern American Farm Homes." *Journal of Interdisciplinary History* 12 (Winter 1982): 409–427.

Porter, Cole. "Don't Fence Me In." In *The Complete Lyrics of Cole Porter*, edited by Robert Kimball. New York: Knopf, 1983.

Proulx, Annie. *Close Range: Wyoming Stories*. New York: Scribner, 1999.

Razac, Olivier. *Barbed Wire: A Political History*. New York: New Press, 2003.

Remarque, Erich Maria. *All Quiet on the Western Front.* Translated by A. W. Wheen. New York: Fawcett Crest, 1975.

Rousseau, Jean-Jacques. *The Major Political Writings of Jean-Jacques Rousseau: The Two "Discourses" and the "Social Contract."* Edited and translated by John T. Scott. 2012. Repr., Chicago: University of Chicago Press, 2014.

Rybczynski, Witold. *One Good Turn: A Natural History of the Screwdriver and the Screw*. New York: Scribner, 2000.

Sandweiss, Martha. "Views and Reviews: Western Art and Western History." In *Under an Open Sky: Rethinking America's Western Past*, edited by William Cronon, George Miles, and Jay Gitlin,185–202. New York: W. W. Norton, 1992.

Schaefer, Jack. *Shane*. Edited by James C. Work. Lincoln: University of Nebraska, 1984.

Schlereth, Thomas. *Material Culture Studies in America: An Anthology Selected, Arranged, & Introduced by Thomas Schlereth*. New York: Altamira Press, 1991.

Seelye, John. "Unlikely Transfusions: *Wise Blood* and *The Grapes of Wrath*." *Steinbeck Studies* 15, no. 1 (2004): 41–45.

Silko, Leslie Marmon. *Ceremony*. New York: Penguin Books, 1977.

Slotkin, Richard. *Regeneration through Violence: The Mythology of the American Frontier, 1600–1860*. Norman: University of Oklahoma Press, 1973.

Stegner, Wallace. "Variations on a Theme by Crèvecoeur." In *New West Reader*, edited by Philip Connors, 105–124. New York: Nation Books, 2005.

Steinbeck, John. *The Grapes of Wrath*. New York: Viking Press, 1939.

———. *To a God Unknown*. New York: Penguin Books, 1995.

St. John, Rachel. *Line in the Sand*. Princeton, NJ: Princeton University Press, 2011.

Szymborska, Wislawa. *Poems, New and Collected*. New York: Mariner Books, 2000.

Thurgood, R. J. *The Complete Encyclopedia of Barbed Wire*. Paducah, KY: Collector Books, 1972.

Turner, Frederick Jackson. "The Significance of the Frontier in American History." In *The Frontier in American History*, edited by Frederick Jackson Turner, 1–38. New York: Henry Holt, 1921.

Tuttle, Jennifer S. "Indigenous Whiteness and Wister's Invisible Indians." In *Reading* The Virginian *in the New West*, edited by Melody Graulich and Stephen Tatum, 89–112. Lincoln: University of Nebraska Press, 2003.

Twain, Mark. *A Connecticut Yankee in King Arthur's Court*. New York: Modern Library, 2001.

Urrea, Luis Alberto. *The Hummingbird's Daughter*. Boston: Little, Brown, 2005.

Vandeburg, C. M. "Barbed Wire." *Wire and Wire Products*. Somerset, NJ, February 1936. Typed transcript in the American Steel & Wire Company Collection. Baker Business Historical Collections. Harvard University.

Vizenor, Gerald Robert. *Manifest Manners: Narratives on Postindian Survivance*. Lincoln: University of Nebraska Press, 1999.

Warren, Arthur G. *Barbed-Wire-—Who Invented It? History of Early Development and Litigation*. N.p., 1926.

———. *History of the Manufacture of Barbed Wire Fencing: Abstract from the Files of the Industrial Museum, American Steel & Wire Company*. Worcester, MA: American Steel & Wire, 1924.

Webb, Walter Prescott. *The Great Plains*. Lincoln: University of Nebraska Press, 1981.

Welch, James. *Winter in the Blood*. New York: Penguin Books, 2008.

Wellington, Jan. "Oscar Wilde's West." *Literary Traveler*, August 19, 2007. http://www.literarytraveler.com/articles/oscar-wildes-west.

Wells, H. G. *The Wonderful Visit*. New York: E. P. Dutton, 1895.

Welter, Barbara. "The Cult of True Womanhood, 1820–1860." *American Quarterly* 18, no. 2 (1966): 151–174.

Wheeler, David L. "The Blizzard of 1886 and Its Effect on the Range Cattle Industry in the Southern Plains." *Southwestern Historical Quarterly* 94, no. 3 (1991): 414–434.

White, Richard. *"It's Your Misfortune and None of My Own": A New History of the American West*. Norman: University of Oklahoma Press, 1991.

Williams, Terry Tempest. *Refuge*. New York: Vintage, 1992.

Wister, Owen. *The Virginian: A Horseman of the Plains*. New York: Macmillan, 1902.

Ziolkowski, Theodore. *Disenchanted Images: A Literary Iconology*. Princeton, NJ: Princeton University Press, 1977.

———. *Fictional Transfigurations of Jesus*. Princeton, NJ: Princeton University Press, 1972.

———. *Varieties of Literary Thematics*. Princeton, NJ: Princeton University Press, 1983.

Films

Barb Wire. Directed by David Hogan. Los Angeles: Propaganda Films and Dark Horse Entertainment, 1996.

Barking Water. Directed by Sterlin Harjo. Tulsa, OK: Indion Production and Dolphin Bay Film, 2009.

HUD. Directed by Martin Ritt. Los Angeles: Paramount Pictures, 1963.

Man without a Star. Directed by King Vidor. Los Angeles: Universal-International, 1955.

The Place of the Falling Waters. Directed by Roy Bigcrane and Thompson Smith. Sacred Land Film Project. Pablo, MT: Salish Kootenai College Media Center, 1990.

Setting a Good Corner (Allegory & Metaphor). Directed by Bruce Nauman. Galisteo, NM: Private Production, 2000.

Shane. Directed by George Stevens. Los Angeles: Paramount Pictures, 1953.
Skins. Directed by Chris Eyre. Los Angeles: First Look Pictures, 2002.
Viva Zapata. Directed by Elia Kazan. Screenplay by John Steinbeck. Los Angeles: Twentieth Century Fox, 1952.

Websites

Amy Clampitt Fund, 1997. http://www.amyclampitt.org/biography/salter.html.
"Barbed Wire Inventors: Jacob Haish," Antique Barbed Wire Society. Accessed March 12, 2017. http://www.antiquebarbedwiresociety.com/inventors_haish.html.
DeKalb, Illinois. Accessed February 20, 2017. http://www.cityofdekalb.com/854/History-of-DeKalb.
"Ein Yahav." Wikipedia, January 22, 2017. https://en.wikipedia.org/wiki/Ein_Yahav.
"From Ireland to Germany to Italy to Mexico: How America's Sources of Immigrants Has Changed in the States, 1850–2013." Pew Research Center, September 28, 2015. http://www.pewhispanic.org/2015/09/28/from-ireland-to-germany-to-italy-to-mexico-how-americas-source-of-immigrants-has-changed-in-the-states-1850-to-2013/.
Gala-Salvador Dalí Foundation. 1937. Accessed March 12, 2017. https://www.salvador-dali.org/en/dali/bio-dali/.
Google Ngram. "Barbed Wire." Accessed March 12, 2017. https://books.google.com/ngrams/graph?content=barbed+wire&year_start=1870&year_end=2000&corpus=15&smoothing=3&share=&direct_url=t1%3B%2Cbarbed%20wire%3B%2Cc0.
The Humane Society of the United States. "Horse Care Guidelines." Accessed March 8, 2017. http://www.humanesociety.org/animals/horses/tips/horse_care_guidelines.html.

INDEX